Transportation Depth Practice Exams
for the Civil PE Exam

Dale R. Gerbetz, PE

PPI2PASS.COM

Professional Publications, Inc. • Belmont, California

Benefit by Registering This Book with PPI

- Get book updates and corrections.
- Hear the latest exam news.
- Obtain exclusive exam tips and strategies.
- Receive special discounts.

Register your book at **ppi2pass.com/register**.

Report Errors and View Corrections for This Book

PPI is grateful to every reader who notifies us of a possible error. Your feedback allows us to improve the quality and accuracy of our products. You can report errata and view corrections at **ppi2pass.com/errata**.

TRANSPORTATION DEPTH PRACTICE EXAMS FOR THE CIVIL PE EXAM

Current release of this edition: 1

Release History

date	edition number	revision number	update
July 2017	1	1	New book.

© 2017 Professional Publications, Inc. All rights reserved.

Printed in the United States of America.

PPI
1250 Fifth Avenue, Belmont, CA 94002
(650) 593-9119
ppi2pass.com

ISBN: 978-1-59126-535-1

Library of Congress Control Number: 2017943720

FEDCBA

Table of Contents

Preface and Acknowledgments

When this book was first being discussed at PPI, an acquisitions editor asked if I was interested in assembling a book of exam-like problems to assist civil transportation examinees. PPI had worked with a small group of dedicated subject matter experts to create, review, and validate problems for a book consisting of practice exams for the transportation depth section of the civil PE exam. The next step was aligning this content with the NCEES exam specifications and creating the practice exams. The problems would support the *Transportation Depth Reference Manual*, which is a companion to Michael R. Lindeburg's *Civil Engineering Reference Manual*. In my circle of civil engineering friends, everyone knows "The CERM." Although there are many books that each of us find helpful and refer to continually, the *Civil Engineering Reference Manual* is the book that we all trust. I was proud to have been approached by PPI, and I was excited to get to work on developing a book that would be as helpful to examinees as the *Civil Engineering Reference Manual* has been to me.

As I reviewed and selected problems for each exam, I strove to be accurate and thorough. I wanted this book to provide the best possible practice for the transportation depth exam. I reviewed the current NCEES specifications for the civil PE exam, returned to the basics of civil and transportation engineering, and carefully considered what civil engineers in the transportation discipline should know after four years in the industry. Success on the NCEES exam requires both theoretical and practical knowledge, so I was confident that drawing on each would be the best approach to a complete review. I also meticulously reviewed the current codes and design standards to ensure this book reflects the NCEES exam specifications at the time of publication.

I would like to thank the original subject matter experts —Alfredo Cely, PE; David Hurwitz, PhD; Gregory Joseph Armstrong; Joshua T. Frohman, PE; Jamie Rana, PE, PTOE; and Eric Donnell, PhD, PE—for their early work in creating, reviewing, and refining problems. Their expertise is evident in the final output, and I thank them for their contribution to the advancement of engineers in our transportation specialty.

Please submit any suspected errors to PPI's errata website at **ppi2pass.com/errata**. All errors will be reviewed, and verified mistakes will be incorporated into future editions of this book. You can also check the latest errata postings to find errors submitted by other readers.

I hope you find this book useful in your preparation for the PE exam. Best of luck to you as you continue your engineering career.

Dale R. Gerbetz, PE

Codes and Standards Used in This Book

The *Civil Engineering Reference Manual* and the *Transportation Depth Reference Manual* are the minimum recommended library for the civil PE transportation depth exam. The exam is based on the following codes as noted by the NCEES transportation specifications.

AASHTO GDHS: *A Policy on Geometric Design of Highways and Streets*, 6th ed., 2011 (including November 2013 errata), American Association of State Highway and Transportation Officials, Washington, DC.

AASHTO GDPS: *Guide for Design of Pavement Structures*, 4th ed., 1993, and 1998 supplement, American Association of State Highway and Transportation Officials, Washington, DC.

AASHTO RDG: *Roadside Design Guide*, 4th ed., 2011 (including February 2012 and July 2015 errata), American Association of State Highway and Transportation Officials, Washington, DC.

AASHTO *Mechanistic-Empirical Pavement Design Guide: A Manual of Practice*, interim ed., July 2008, American Association of State Highway and Transportation Officials, Washington, DC.

AASHTO *Guide for the Planning, Design, and Operation of Pedestrian Facilities*, 1st ed., 2004, American Association of State Highway and Transportation Officials, Washington, DC.

AASHTO HSM: *Highway Safety Manual*, 1st ed., 2010, vols. 1–3 (including February 2012 errata), American Association of State Highway and Transportation Officials, Washington, DC.

AI: *The Asphalt Handbook* (MS-4), 7th ed., 2007, Asphalt Institute, Lexington, KY.

HCM: *Highway Capacity Manual*, 2010 ed., vols. 1–3. Transportation Research Board, National Research Council, Washington, DC. This includes the following:

> Approved HCM 2010 Corrections and Clarifications (as of January 2014)
>
> Approved HCM 2010 Interpretations (as of January 2014)
>
> Replacement HCM 2010 Volume 1–3 pages (April 2014)
>
> Replacement HCM 2010 Volume 1–3 pages (January 12–February 13)
>
> Replacement HCM 2010 Volume 1–3 pages (March 2013)

MUTCD: *Manual on Uniform Traffic Control Devices*, 2009, including Revisions 1 and 2, dated May 2012, U.S. Department of Transportation-Federal Highway Administration, Washington, DC.

PCA: *Design and Control of Concrete Mixtures*, 15th ed., 2011, Portland Cement Association, Skokie, IL.

FHWA: *Hydraulic Design of Highway Culverts*, Hydraulic Design Series Number 5, Publication No. FHWA-HIF-12-026, 3rd ed., April 2012, U.S. Department of Transportation, Federal Highway Administration, Washington, DC.

Introduction

ABOUT THE BOOK

Transportation Depth Practice Exams for the Civil PE Exam includes two exams designed to match the format and specifications of the transportation depth section of the civil PE exam. Like the actual exam, the exams in this book contain 40 multiple-choice problems, and each problem takes an average of six minutes to solve. Most of the problems are quantitative, requiring calculations to arrive at the correct answer. A few are nonquantitative.

Each problem has four answer options—labeled A, B, C, and D—one of which is correct (or "most nearly correct"). Numerical answer options are displayed in increasing value. Incorrect answer options may consist of one or more "logical distractors," the term used by NCEES to designate options that may seem correct. Incorrect options represent answers found by making common mistakes. These may be simple mathematical errors, such as failing to square a term in an equation, or more serious errors, such as using the wrong equation.

The solutions in this book are presented step-by-step to help you follow the logical development of the solving approach and to provide examples of how you may want to solve similar problems on the exam.

Solutions presented for each problem may represent only one of several methods for obtaining the correct answer. Alternative problem-solving methods may also produce correct answers.

ABOUT THE EXAM

The Principles and Practice of Engineering (PE) exam is administered by the National Council of Examiners for Engineering and Surveying (NCEES). The civil PE exam is an eight-hour exam divided into a four-hour morning breadth exam and a four-hour afternoon depth exam. The morning breadth exam consists of 40 multiple-choice problems covering eight areas of general civil engineering knowledge: project planning; means and methods; soil mechanics; structural mechanics; hydraulics and hydrology; geometrics; materials; and site development. As the "breadth" designation implies, morning exam problems are general in nature and wide-ranging in scope. All examinees take the same breadth exam.

For the afternoon depth exam, you must select a depth section from one of the five subdisciplines: construction, geotechnical, structural, transportation, or water

resources and environmental. The problems on the afternoon depth exam require more specialized knowledge than those on the morning breadth exam. Topics and the distribution of problems on the transportation depth exam are as follows.

- **Traffic Engineering (Capacity Analysis and Transportation Planning) (11 questions)**

  uninterrupted flow (e.g., level of service, capacity); street segment interrupted flow (e.g., level of service, running time, travel speed); intersection capacity (e.g., at grade, signalized, roundabout, interchange); traffic analysis (e.g., volume studies, peak hour factor, speed studies, modal split); trip generation and traffic impact studies; accident analysis (e.g., conflict analysis, accident rates, collision diagrams); nonmotorized facilities (e.g., pedestrian, bicycle); traffic forecast; highway safety analysis (e.g., crash modification factors, *Highway Safety Manual*)

- **Horizontal Design (4 questions)**

  basic curve elements (e.g., middle ordinate, length, chord, radius); sight distance considerations; superelevation (e.g., rate, transitions, method, components); special horizontal curves (e.g., compound/reverse curves, curve widening, coordination with vertical geometry)

- **Vertical Design (4 questions)**

  vertical curve geometry; stopping and passing sight distance (e.g., crest curve, sag curve); vertical clearance

- **Intersection Geometry (4 questions)**

  intersection sight distance; interchanges (e.g., freeway merge, entrance and exit design, horizontal design, vertical design); at-grade intersection layout, including roundabouts

- **Roadside and Cross-Section Design (4 questions)**

  forgiving roadside concepts (e.g., clear zone, recoverable slopes, roadside obstacles); barrier design (e.g., barrier types, end treatments, crash cushions); cross-section elements (e.g., lane widths, shoulders, bike lane, sidewalks); Americans with Disabilities Act (ADA) design considerations

- **Signal Design (3 questions)**

  signal timing (e.g., clearance intervals, phasing, pedestrian crossing timing, railroad preemption); signal warrants

- **Traffic Control Design (3 questions)**

  signs and pavement markings; temporary traffic control

- **Geotechnical and Pavement (4 questions)**

  sampling and testing (e.g., subgrade resilient modulus, CBR, R-Values, field tests); soil stabilization techniques, settlement and compaction, excavation, embankment, and mass balance; design traffic analysis and pavement design procedures (e.g., flexible and rigid pavement); pavement evaluation and maintenance measures (e.g., skid, roughness, rehabilitation treatments)

- **Drainage (2 questions)**

  hydrology (e.g., Rational method, hydrographs, SCS/NRCS method), including runoff detention/retention/water quality mitigation measures; hydraulics, including culvert and stormwater collection system design (e.g., inlet capacities, pipe flow, hydraulic energy dissipation), and open-channel flow

- **Alternatives Analysis (1 question)**

  economic analysis (e.g., present worth, lifecycle costs)

HOW TO USE THIS BOOK

Prior to taking these practice exams, locate and organize relevant resources and materials as if you are taking the actual exam. Refer to the Codes and Standards Used in This Book section for guidance on materials. Also, visit **ppi2pass.com/stateboards** for a link to your state's board of engineering, and check for any state-specific restrictions on materials you are allowed to bring to the exam. You should also check NCEES' calculator policy at **ppi2pass.com/calculators** to ensure your calculator can be used on the exam.

The two exams in this book allow you to structure your own exam preparation in the way that is best for you. For example, you might choose to take one exam as a pretest to assess your knowledge and determine the areas in which you need more review, and then take the second after you have completed additional study. Alternatively, you might choose to use one exam as a guide for how to solve different types of problems, reading each problem and solution in kind, and then use the second exam to evaluate what you learned.

Whatever your preferred exam preparation method, these exams will be most useful if you restrict yourself to exam-like conditions when solving the problems. When you are ready to begin an exam, set a timer for four hours. Use only the calculator and references you have gathered for use on the exam. Use the space provided near each problem for your calculations, and mark your answer on the answer sheet.

When you finish taking an exam, check your answers against the answer key to assess your performance. Review the solutions to any problems you answered incorrectly or were unable to answer. Read the author commentaries for tips, and compare your problem-solving approaches against those given in the solutions.

Nomenclature

PPI ● ppi2pass.com

Nomenclature

a	layer coefficient	in	cm
A	area	ft^2	m^2
A	volume	ac-ft	–
A	autos per household	veh/house	–
A	change in gradient	–	–
A	annual cost	–	–
A_p	pedestrian space	ft^2/ped	m^2/ped
ADT	average daily traffic	veh/day	veh/d
b	bottom width	ft	m
b	base (of a triangle)	min	min
BCR	benefit cost ratio	–	–
c	capacity	pcph	pcph
c	commute cost	\$	\$
C	cycle	sec	s
C	runoff coefficient	–	–
C_e	existing conditions runoff coefficient	–	–
C_r	composite coefficient	–	–
CZ	clear zone	ft	m
C_{ZC}	minimum clear zone distance	ft	m
d	control delay	sec	s
d_1	uniform control delay	sec	s
d_2	incremental delay	sec	s
d_3	initial queue delay	sec	s
d	distance	ft	m
D	degree of curve	deg	deg
D	density	veh/mi	veh/km
D	pavement	in	mm
D	duration	sec	s
D_p	phase of duration	sec	s
e	superelevation rate	ft/ft	m/m
E	elevation	ft	m
E_B	base resilient modulus	lbf/in^2	Pa
f	adjustment factor	–	–
f	friction factor	–	–
f_s	side friction factor	–	–
F	salvage value	–	–
g	green time	sec	s
g_e	green extension time	sec	s
g_s	green service time	sec	s
G	grade	–	–
G	specific gravity	–	–

G	subsequent year increase	–	–
G_{mm}	maximum specific gravity	–	–
h	height	ft	m
HSO	horizontal sightline offset	ft	m
G_b	specific gravity of asphalt	–	–
G_{se}	specific gravity of aggregate	–	–
i	intensity	in/hr	cm/h
I	angle	deg	deg
ISD	sight distance	ft	m
K_{CZ}	horizontal curve adjustment factor	–	–
l_1	start-up lost time	sec	s
L	length	ft	m
L_a	lateral extent of area	ft	m
L_c	clear zone distance	ft	m
L_t	runout	ft	m
m	drainage coefficient	–	–
n	Manning's roughness coefficient	–	–
N	number	–	–
p	probability	–	–
P	persons per household	–	–
P	present worth, purchase price	–	–
P	perimeter	ft	m
P_b	percent asphalt	–	–
P_s	percent aggregate	–	–
P_w	wetted perimeter	ft	m
P	fractional value	–	–
PC	point of curvature	sta	sta
PF	progression factor	–	–
PHF	peak hour factor	–	–
PV	present value	–	–
PT	point of tangency	sta	sta
Q	flow	ft^3/sec	m^3/s
Q_r	maximum queued vehicles	veh	veh
R	hydraulic radius	ft	m
R	rate	various	various
R_C	red clearance interval	sec	s
s	saturation flow rate	veh/hr	veh/h
s	braking/skidding distance	ft	m
S	speed	ft/sec	m/s
S	sight stopping distance	ft	m
S	slope	ft/ft	m/m

S	sight distance	ft	m
SN	structural number	–	–
SSD	stopping sight distance	ft	m
t	thickness	in	mm
t	time	sec	s
t_e	existing time of concentration	min	min
t_g	time gap	sec	s
t_p	proposed time of concentration	min	min
T	gutter spread	ft	m
T	number of household trips	–	–
T	tangent length	ft	m
v	velocity	mi/hr	km/h
v	volume	ft^3	m^3
v_p	maximum flow rate	pcphpl	pcphpl
V	volume (of traffic)	veh	veh
W	wait time	min	min
W	width	ft	m
W_e	effective width	ft	m
X	length-of-need	ft	m
Y	degree of saturation	deg	deg
Y	lateral offset	–	–
Y	yellow change interval	sec	s
BVC	beginning of vertical curve	sta	sta
HP	high point	ft	m

Symbols

ρ	density	lbm/ft^3

Subscripts

BVC	beginning of vertical curve
G	gradient cost
HP	high point
HV	heavy vehicle
L	lane
p	driver population
p	pedestrian
P/A	present annual worth
PC	point of curvature
PI	point of intersection
PT	point of tangency
R	runoff
R	runout
ROW	right-of-way

Practice Exam 1 Instructions

In accordance with the rules established by your state, you may use textbooks, handbooks, bound reference materials, and any approved battery- or solar-powered, silent calculator to work this examination. However, no blank papers, writing tablets, unbound scratch paper, or loose notes are permitted. Sufficient room for scratch work is provided in the Examination Booklet.

You are not permitted to share or exchange materials with other examinees. However, the books and other resources used in this afternoon session do not have to be the same as were used in the morning session.

You will have four hours in which to work this session of the examination. Your score will be determined by the number of questions that you answer correctly. There is a total of 40 questions. All 40 questions must be worked correctly in order to receive full credit on the exam. There are no optional questions. Each question is worth 1 point. The maximum possible score for this section of the examination is 40 points.

Partial credit is not available. No credit will be given for methodology, assumptions, or work written in your Examination Booklet.

Record all of your answers on the Answer Sheet. No credit will be given for answers marked in the Examination Booklet. Mark your answers with the official examination pencil provided to you. Marks must be dark and must completely fill the bubbles. Record only one answer per question. If you mark more than one answer, you will not receive credit for the question. If you change an answer, be sure the old bubble is erased completely; incomplete erasures may be misinterpreted as answers.

If you finish early, check your work and make sure that you have followed all instructions. After checking your answers, you may turn in your Examination Booklet and Answer Sheet and leave the examination room. Once you leave, you will not be permitted to return to work or change your answers.

When permission has been given by your proctor, break the seal on the Examination Booklet. Check that all pages are present and legible. If any part of your Examination Booklet is missing, your proctor will issue you a new Booklet.

Principles and Practice of Engineering Examination

Afternoon Session
Practice Exam 1

1. Ⓐ Ⓑ Ⓒ Ⓓ
2. Ⓐ Ⓑ Ⓒ Ⓓ
3. Ⓐ Ⓑ Ⓒ Ⓓ
4. Ⓐ Ⓑ Ⓒ Ⓓ
5. Ⓐ Ⓑ Ⓒ Ⓓ
6. Ⓐ Ⓑ Ⓒ Ⓓ
7. Ⓐ Ⓑ Ⓒ Ⓓ
8. Ⓐ Ⓑ Ⓒ Ⓓ
9. Ⓐ Ⓑ Ⓒ Ⓓ
10. Ⓐ Ⓑ Ⓒ Ⓓ

11. Ⓐ Ⓑ Ⓒ Ⓓ
12. Ⓐ Ⓑ Ⓒ Ⓓ
13. Ⓐ Ⓑ Ⓒ Ⓓ
14. Ⓐ Ⓑ Ⓒ Ⓓ
15. Ⓐ Ⓑ Ⓒ Ⓓ
16. Ⓐ Ⓑ Ⓒ Ⓓ
17. Ⓐ Ⓑ Ⓒ Ⓓ
18. Ⓐ Ⓑ Ⓒ Ⓓ
19. Ⓐ Ⓑ Ⓒ Ⓓ
20. Ⓐ Ⓑ Ⓒ Ⓓ

21. Ⓐ Ⓑ Ⓒ Ⓓ
22. Ⓐ Ⓑ Ⓒ Ⓓ
23. Ⓐ Ⓑ Ⓒ Ⓓ
24. Ⓐ Ⓑ Ⓒ Ⓓ
25. Ⓐ Ⓑ Ⓒ Ⓓ
26. Ⓐ Ⓑ Ⓒ Ⓓ
27. Ⓐ Ⓑ Ⓒ Ⓓ
28. Ⓐ Ⓑ Ⓒ Ⓓ
29. Ⓐ Ⓑ Ⓒ Ⓓ
30. Ⓐ Ⓑ Ⓒ Ⓓ

31. Ⓐ Ⓑ Ⓒ Ⓓ
32. Ⓐ Ⓑ Ⓒ Ⓓ
33. Ⓐ Ⓑ Ⓒ Ⓓ
34. Ⓐ Ⓑ Ⓒ Ⓓ
35. Ⓐ Ⓑ Ⓒ Ⓓ
36. Ⓐ Ⓑ Ⓒ Ⓓ
37. Ⓐ Ⓑ Ⓒ Ⓓ
38. Ⓐ Ⓑ Ⓒ Ⓓ
39. Ⓐ Ⓑ Ⓒ Ⓓ
40. Ⓐ Ⓑ Ⓒ Ⓓ

Practice Exam 1

1. A 1 mi length of highway is widened with two additional 12 ft lanes. The highway is located on a closed watershed. The highway widening has no impact on residential land uses. The increase of impervious areas will decrease the existing time of concentration by 20%. Data for rainfall intensity, duration, and frequency is shown.

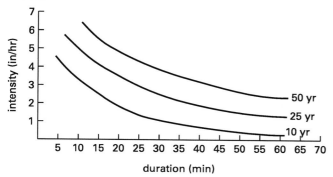

existing land use	area (ac)	runoff coefficient, C
meadow	5.2	0.40
residential	10.3	0.50
highway	3.1	0.90

existing conditions coefficient, $C_e = 0.54$

existing time of concentration, $t_e = 25$ min

For a 10-year storm event, the increase in peak runoff due to roadway widening is most nearly

(A) 5 ft³/sec

(B) 7 ft³/sec

(C) 14 ft³/sec

(D) 21 ft³/sec

2. A single-unit truck on a minor road turns left onto an arterial roadway at an intersection. Only the minor road has stop signs. The arterial roadway has two lanes in each direction, a median that is 24 ft wide, and a posted speed limit of 45 mph. The roadways intersect at a flat grade. Most nearly, what is the minimum required intersection sight distance for the turn?

(A) 545 ft

(B) 630 ft

(C) 770 ft

(D) 815 ft

3. Cross section cut and fill areas for a corridor model output are shown.

station	cut area (yd²)	fill area (yd²)
10+00	10	20
11+00	1	30
12+00	5	10
13+00	6	6
14+00	10	5
15+00	12	8
16+00	10	7
17+00	20	2
18+00	20	2
19+00	30	2

Beginning from sta 10+00, the first balanced station point occurs at

(A) sta 13+00

(B) sta 14+00

(C) sta 18+00

(D) sta 19+00

4. A pavement section is designed for a reliability of 99% with a standard deviation of 0.2 and a design serviceability loss of 1. The pavement section has a 5 in asphaltic concrete surface course with a layer strength coefficient of 0.44/in, an 8 in crushed concrete base with a layer strength coefficient of 0.12/in and a drainage coefficient of 1.0, and a 12 in stabilized subgrade with a layer strength coefficient of 0.07/in and a drainage coefficient of 1.0. The resilient modulus for the subgrade is 15 kips/in².

According to the AASHTO *Guide for Design of Pavement Structures*, the design equivalent single-axle load (ESAL) is most nearly

- (A) 2.0×10^6
- (B) 6.0×10^6
- (C) 10.0×10^6
- (D) 15.0×10^6

5. The roadway for an urban collector road is to be rehabilitated. The design for the original roadway is shown.

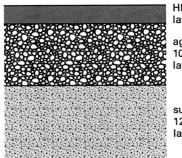

HMA, 4 in
layer coefficient $a_1 = 0.4$/in

aggregate base
10 in
layer coefficient $a_2 = 0.2$/in

subgrade
12 in
layer coefficient $a_3 = 0.1$/in

Due to changes in traffic volume, the thickness of the asphalt for the rehabilitated roadway can be less than the original. The aggregate base and subgrade have a drainage coefficient equal to 1.0.

The structural number for the rehabilitated asphalt layer is 4.4. The minimum asphalt layer thickness is

- (A) 2.0 in
- (B) 2.5 in
- (C) 3.0 in
- (D) 4.0 in

6. Lighting is added to a 12 ft wide paved bicycle path which is designed for a bicycle approach speed of 20 mph. The lighting fixtures are mounted to square posts that encroach 1 ft onto the paved bicycle path as shown.

Obstruction pavement markings are added to warn cyclists of the encroaching light posts. Most nearly, what is the taper length requirement, L, for the obstruction pavement markings?

- (A) 5 ft
- (B) 10 ft
- (C) 20 ft
- (D) 40 ft

7. A roadway profile with storm drainage peak flows for each subbasin of an inlet is shown.

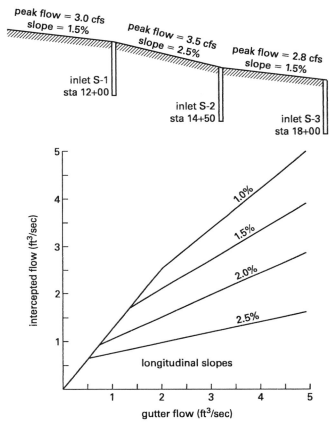

The roadway cross slope is 2%, and the Manning's roughness coefficient for the concrete gutter is 0.012. Most nearly, what is the maximum gutter spread between sta 14+50 and sta 18+00?

- (A) 9 ft
- (B) 11 ft
- (C) 12 ft
- (D) 15 ft

8. A rural collector road is under construction. The subgrade of the pavement layer must have a minimum relative compaction of 98%. Proctor test results from subgrade soil samples are summarized in the graph.

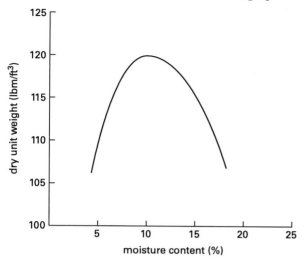

Samples for soil density and moisture content tests were collected at four locations.

sample	density (lbm/ft^3)	percent moisture content
1	120	6
2	120	17
3	125	5
4	125	13

Which sample is above 98% compaction?

(A) 1

(B) 2

(C) 3

(D) 4

9. A traffic analysis zone (TAZ) contains 425 houses with an average of 3 persons and 2.2 vehicles per household. For the TAZ, the number of household trips per day, T, is a function of the number of persons per household, P, and the number of autos per household, A.

$$T = 0.78 + 1.6P + 2.4A$$

The average number of trips per day within the TAZ is most nearly

(A) 900

(B) 1300

(C) 2800

(D) 4600

10. According to the *Highway Capacity Manual* (HCM), a Class II two-lane highway operating under uninterrupted flow conditions serves

   I.   as a scenic or recreational route

   II.   moderately developed areas

   III.   mostly relatively short trips where motorists do not expect to travel at high speeds

   IV.   as a daily commuter route or primary connector of major traffic generators

(A) I and II

(B) I and III

(C) II and IV

(D) III and IV

11. A roundabout is being designed with an entry design speed of 25 mph and an inscribed circle diameter of 120 ft to 140 ft. The expected typical daily volume is 17,000 to 19,000 vehicles per day. According to the AASHTO *Green Book*, what type of roundabout is this?

(A) urban

(B) mini

(C) single-lane

(D) multilane

12. A vehicle travelling on a level road brakes, and its wheels lock up. The vehicle skids across asphalt and grass before crashing into a tree at 25 mph. The crash report diagram is shown.

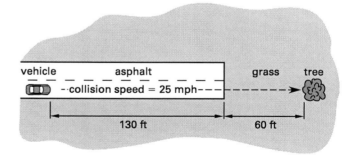

The vehicle begins braking 190 ft away from the crash site. The coefficients of dynamic friction between the tires and asphalt and between the tires and grass are 0.8 and 0.5, respectively. Most nearly, what is the speed of the vehicle when braking begins?

(A) 66 mph

(B) 68 mph

(C) 70 mph

(D) 72 mph

13. A car travels at 45 mph up a crest vertical curve that is 900 ft in length. The curve has an algebraic grade difference of 3.25%. The driver's eyes are 3.5 ft above the road surface and are focused on an object 0.50 ft above the road surface. The object is in the road directly ahead of the driver. Most nearly, what is the stopping sight distance?

(A) 470 ft

(B) 530 ft

(C) 610 ft

(D) 660 ft

14. A four-leg intersection is designed to have a single shared lane for right, through, and left traffic on each approach. The design speed of the major roadway is 45 mph. The minor road has a downgrade of 4% toward the intersection. Controls are on the minor road.

A stopped combination truck on the minor road turns right.

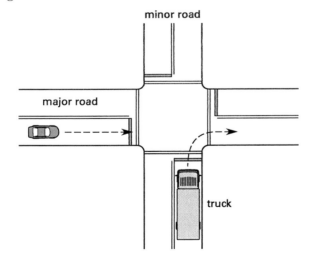

According to the AASHTO *Green Book*, what is most nearly the minimum sight distance along the major roadway needed for the truck to safely turn right?

(A) 430 ft

(B) 589 ft

(C) 695 ft

(D) 721 ft

15. A roadway segment has a diamond interchange with three eastbound approaching lanes. The origin-destination demands are shown.

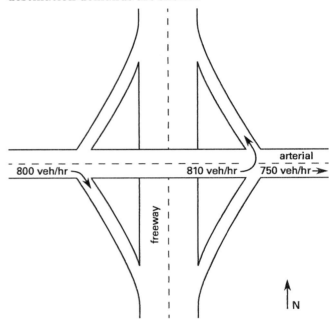

According to the *Highway Capacity Manual* (HCM), the eastbound leftmost lane is most nearly what percentage of total traffic volume?

(A) 24%

(B) 27%

(C) 33%

(D) 41%

16. A roadway with a horizontal curve deflecting to the left is shown.

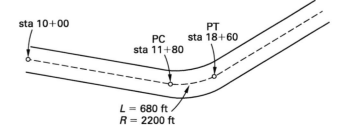

The roadway is a two-way, two-lane rural highway with a design speed of 60 mph, a lane width of 12 ft, and a maximum superelevation rate of 10%. The normal crown of the road is 2%.

Most nearly, what are the full superelevation, runoff, and runout values for the roadway, respectively?

(A) 7%, 184 ft, and 53 ft

(B) 7%, 53 ft, and 187 ft

(C) 9%, 184 ft, and 53 ft

(D) 8%, 53 ft, and 184 ft

17. The installation of a traffic control signal at an intersection is justified by Warrant 9 of the *Manual on Uniform Traffic Control Devices* (MUTCD). The LEAST likely thing to find at the intersection is

(A) a grade crossing with automatic gates

(B) a grade crossing with flashing-light signals

(C) preemption control for emergency vehicles

(D) a minor street with an actuated traffic control signal

18. A roadway is repaved with pavement that is to provide 10 years of useful life. Annual maintenance costs for the new pavement accrue to $1000 the first year and increase by $250 each year for every subsequent year. The interest rate is 7% and is compounded annually. Most nearly, what is the present worth of the maintenance costs over the lifetime of the new pavement?

(A) $12,500

(B) $14,000

(C) $17,500

(D) $19,000

19. According to the AASHTO *Roadside Design Guide* (RDG), the likelihood of automobile crashes involving utility poles is minimized by treating several utility poles along a corridor. Which of the following is NOT a strategy for treating utility poles?

(A) using breakaway poles

(B) placing utilities underground

(C) relocating poles farther from the roadway

(D) decreasing the number of poles along the corridor

20. A guardrail barrier is being designed to reduce the likelihood of crashes with a tree. The tree is located within the clear zone as shown in Fig. 5-39 from the AASHTO *Roadside Design Guide* (RDG).

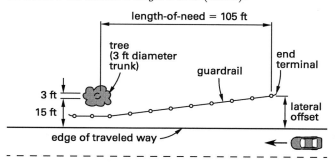

(not to scale)

The design speed is 60 mph and the traffic volume is 5700 vehicles per day. Most nearly, what is the lateral offset of the guardrail?

(A) 9 ft

(B) 10 ft

(C) 15 ft

(D) 18 ft

21. A proposal to widen an existing roadway is under way. The cross-sectional design for the proposal is shown.

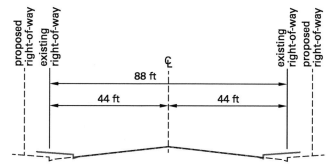

The existing roadway is classified as a minor arterial street. After widening, the roadway will be a divided arterial street. The design parameters for the proposal include a median that is 18 ft wide; two striped travel lanes, one 14 ft wide and one 12 ft wide; a striped bike lane 8 ft wide; lane stripes measuring 8 in wide; curbs 8 in high; gutters 2 ft wide and adjacent to the bike lane; sidewalks 6 ft wide and adjacent to the curb on each side of the street; and clear zones measuring 12 ft wide from curb line to right-of-way.

The total existing right-of-way is 88 ft wide with 44 ft on each side of the street's centerline. Most nearly, how much additional right-of-way is needed on each side of the street to fit the design parameters?

(A) 11 ft

(B) 13 ft

(C) 15 ft

(D) 21 ft

22. An urban street with signalized intersections has a length of 0.5 mi. The street has a free flow speed of 40 mph. There are six influential access points, and the average delay per influential access point is 22 sec. The proximity adjustment factor is 1.01.

According to the *Highway Capacity Manual* (HCM), the segment running time is most nearly

(A) 180 sec

(B) 195 sec

(C) 210 sec

(D) 225 sec

23. According to the design requirements in the Americans with Disabilities Act (ADA), a curb ramp is to have a ramp slope no greater than 12H:1V, wing slopes no greater than 10H:1V, and a 4 ft minimum landing at the top of the ramp at 2% maximum slope.

An illustration for a curb ramp is shown.

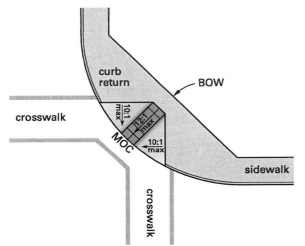

The curb has a height of 8 in. The elevation difference longitudinally across the ramp is equal to the curb

height. Most nearly, what is the minimum horizontal distance from middle of curve (MOC) perpendicular to the back of walk (BOW)?

(A) 8 ft

(B) 10 ft

(C) 12 ft

(D) 14 ft

24. A second circulating lane is added to a roundabout with a single entry lane and a single circulating lane. The single-lane roundabout has a lane capacity of 460 passenger cars per hour. According to the *Highway Capacity Manual* (HCM), the new lane capacity after adding the second circulating lane is most nearly

(A) 460 pcph

(B) 575 pcph

(C) 600 pcph

(D) 920 pcph

25. The horizontal curve of a roadway section has a 5000 ft radius. The tangent length is 600 ft, and the point of intersection is located at station 204+00. The station at the point of tangency is

(A) 198+00

(B) 203+97

(C) 209+00

(D) 209+96

26. A section of a rural two-lane highway with 12 ft lanes has a 9° curve. At the inside of the curve, a ground-mounted billboard that is 50 ft in length is in close proximity to a right-of-way line. The center of the billboard is located 77 ft from the centerline of the highway. Most nearly, what is the minimum sight distance along the curve?

(A) 232 ft

(B) 487 ft

(C) 520 ft

(D) 601 ft

27. Two parallel roadways are being connected with a new reversed curve walkway consisting of two simple circular curves. The roadways are 1640 ft apart. The radius of the first curve is 1300 ft, and the radius of the second curve is 1970 ft. Most nearly, what is the length of the new walkway?

(A) 1500 ft

(B) 2063 ft

(C) 3430 ft

(D) 4521 ft

28. A crest vertical curve is 820 ft long and has a positive 3.5% grade and a negative 2.0% grade. The grades intersect at sta 142+70 at an elevation of 625 ft. Most nearly, what is the station and elevation of the high point of the curve?

(A) 138+60; 611 ft

(B) 143+82; 522 ft

(C) 143+82; 620 ft

(D) 153+80; 620 ft

29. A sag vertical curve is designed for a speed of 35 mph. The grade into the curve is negative 2.5%, and the grade out of the curve is positive 3.7%. Most nearly, what is the difference between the minimum curve length required for stopping sight distance and the minimum curve length required for rider comfort?

(A) 126 ft

(B) 130 ft

(C) 140 ft

(D) 163 ft

30. A railroad overpass perpendicularly traverses a roadway section. The roadway section has a sag vertical curve that is 300 ft in length. The curve's incoming grade of −1.3% intersects the outgoing grade of 1.8% at sta 74+25 and elevation 810 ft. The overpass must be at least 18 ft above the lowest point on the curve. What is most nearly the minimum elevation of the overpass?

(A) 816 ft

(B) 819 ft

(C) 829 ft

(D) 833 ft

31. A pre-timed traffic signal is located at an intersection several miles from the next adjacent traffic signal. The intersection has a saturation flow of 1900 vehicles per hour per lane. The pre-timed signal has characteristics as shown in the table.

phase	demand (veh/hr)	yellow change interval (sec)	red clearance interval (sec)
1	100	4.5	3.0
2	350	5.0	4.0
3	250	4.5	4.5

The optimum cycle length for the intersection is most nearly

(A) 60 sec

(B) 70 sec

(C) 75 sec

(D) 80 sec

32. Between mile marker 110 and mile marker 120 of a two-lane roadway, there are 16 crashes over a two-year period. The average daily traffic (ADT) for the roadway segment is 12,000 vehicles per day. What is the accident rate per 100 million vehicles miles (HMVM) for the roadway segment?

(A) 16.9 crashes per HMVM

(B) 17.5 crashes per HMVM

(C) 18.3 crashes per HMVM

(D) 20.3 crashes per HMVM

33. The peak morning and afternoon hours for mid-block crossings at a roadway location are shown.

time period	approach volume (southbound)	approach volume (northbound)	pedestrian volume
morning			
7:00–7:15	211	125	12
7:15–7:30	216	124	16
7:30–7:45	204	116	33
7:45–8:00	199	128	36
afternoon			
4:00–4:15	102	198	12
4:15–4:30	122	199	10
4:30–4:45	132	208	16
4:45–5:00	130	200	11

The roadway has a posted speed limit of 40 mph. The closest traffic signal is 330 ft away from the midblock location. When is the pedestrian volume warrant satisfied?

(A) peak morning

(B) peak afternoon

(C) both peak morning and peak afternoon

(D) neither peak morning nor peak afternoon

34. A vehicle is traveling on a roadway with a negative 6% grade. When the vehicle brakes, its wheels lock up. The vehicle skids and crashes into a concrete bridge overpass support. The skid marks measure 517 ft in length. The vehicle was traveling 30 mph upon impact. The posted speed limit on the roadway is 60 mph. The coefficient of friction between the tires and road is 0.30.

How many miles per hour over the posted speed limit was the vehicle traveling when its wheels locked up?

(A) 1 mph

(B) 8 mph

(C) 11 mph

(D) 16 mph

35. An urban freeway segment on rolling terrain is designed to carry a commuter volume of 5000 vehicles per hour with a distribution of 90% passenger cars and 10% trucks. The peak hour factor is 0.92. The free flow speed is 65 mph. The number of lanes required for level of service (LOS) C is

(A) 2

(B) 3

(C) 4

(D) 5

36. A six-lane urban freeway with three lanes in each direction is on rolling terrain. The directional peak hour volume is 3250 vehicles per hour. Traffic includes 5% trucks, 2% buses, and 2% recreational vehicles (RV). The free flow speed is 55 mph. The driver population adjustment factor is 0.95. The peak hour factor is 0.87. The level of service (LOS) is most nearly

(A) LOS B

(B) LOS C

(C) LOS D

(D) LOS E

37. A pedestrian walkway outside an assisted living facility is 10 ft wide with a decorative planting area 3 ft wide. The peak pedestrian flow rate is estimated at 25 pedestrians per 15 min. According to the *Highway Capacity Manual* (HCM), what is the anticipated pedestrian peak level of service (LOS)?

(A) LOS A

(B) LOS B

(C) LOS C

(D) LOS D

38. A merging taper is being designed for the transition and termination areas of a single temporary lane closure. The closure is on a divided multilane highway with lanes that are 12 ft wide. The highway has an off-peak 85th percentile speed of 55 mph. According to the *Manual on Uniform Traffic Control Devices* (MUTCD), the minimum taper length recommended is

(A) 303 ft

(B) 330 ft

(C) 484 ft

(D) 660 ft

39. A temporary traffic control zone uses flagger control during the daytime. According to the *Manual on Uniform Traffic Control Devices* (MUTCD), which of the following is NOT a required element for the STOP/SLOW paddle?

(A) The paddle must have an octagonal shape.

(B) The paddle must be reflectorized.

(C) The STOP face must have white letters and a white border on a red background.

(D) The SLOW face must have black letters and a black border on an orange background.

40. A traffic study is conducted at a standard four-leg intersection with one shared travel lane for left, through, and right directions on each approach. Count data for two hours of turning movement is shown in the table.

interval	from north			from south			from east			from west		
	left	through	right	left	through	right	left	through	right	left	through	right
1:00–1:15	5	17	11	11	30	15	3	5	0	8	12	2
1:15–1:30	5	16	12	12	36	16	4	6	0	12	11	3
1:30–1:45	7	18	14	14	42	18	2	7	2	11	16	1
1:45–2:00	8	20	11	11	58	16	4	5	0	15	18	5
2:00–2:15	12	37	9	9	45	20	5	9	0	18	19	2
2:15–2:30	11	40	12	12	49	17	7	12	1	6	20	6
2:30–2:45	14	35	15	15	48	15	8	8	0	4	11	3
2:45–3:00	12	30	16	16	46	15	9	9	0	5	16	5

According to the *Highway Capacity Manual* (HCM), what are most nearly the peak hour and peak hour factor for the intersection during this two-hour period?

(A)   1:45–2:45; 0.93

(B)   1:45–2:45; 0.96

(C)   2:00–3:00; 0.95

(D)   2:00–3:00; 0.98

STOP!

DO NOT CONTINUE!

This concludes the Afternoon Session of the examination. If you finish early, check your work and make sure that you have followed all instructions. After checking your answers, you may turn in your examination booklet and answer sheet and leave the examination room. Once you leave, you will not be permitted to return to work or change your answers.

Practice Exam 2 Instructions

In accordance with the rules established by your state, you may use textbooks, handbooks, bound reference materials, and any approved battery- or solar-powered, silent calculator to work this examination. However, no blank papers, writing tablets, unbound scratch paper, or loose notes are permitted. Sufficient room for scratch work is provided in the Examination Booklet.

You are not permitted to share or exchange materials with other examinees. However, the books and other resources used in this afternoon session do not have to be the same as were used in the morning session.

You will have four hours in which to work this session of the examination. Your score will be determined by the number of questions that you answer correctly. There is a total of 40 questions. All 40 questions must be worked correctly in order to receive full credit on the exam. There are no optional questions. Each question is worth 1 point. The maximum possible score for this section of the examination is 40 points.

Partial credit is not available. No credit will be given for methodology, assumptions, or work written in your Examination Booklet.

Record all of your answers on the Answer Sheet. No credit will be given for answers marked in the Examination Booklet. Mark your answers with the official examination pencil provided to you. Answers marked in pen may not be graded correctly. Marks must be dark and must completely fill the bubbles. Record only one answer per question. If you mark more than one answer, you will not receive credit for the question. If you change an answer, be sure the old bubble is erased completely; incomplete erasures may be misinterpreted as answers.

If you finish early, check your work and make sure that you have followed all instructions. After checking your answers, you may turn in your Examination Booklet and Answer Sheet and leave the examination room. Once you leave, you will not be permitted to return to work or change your answers.

When permission has been given by your proctor, break the seal on the Examination Booklet. Check that all pages are present and legible. If any part of your Examination Booklet is missing, your proctor will issue you a new Booklet.

WAIT FOR PERMISSION TO BEGIN

Name: _____
          Last   First   Middle Initial

Examinee number: _____

Examination Booklet number: _____

Principles and Practice of Engineering Examination

Afternoon Session
Practice Exam 2

41. (A) (B) (C) (D)
42. (A) (B) (C) (D)
43. (A) (B) (C) (D)
44. (A) (B) (C) (D)
45. (A) (B) (C) (D)
46. (A) (B) (C) (D)
47. (A) (B) (C) (D)
48. (A) (B) (C) (D)
49. (A) (B) (C) (D)
50. (A) (B) (C) (D)

51. (A) (B) (C) (D)
52. (A) (B) (C) (D)
53. (A) (B) (C) (D)
54. (A) (B) (C) (D)
55. (A) (B) (C) (D)
56. (A) (B) (C) (D)
57. (A) (B) (C) (D)
58. (A) (B) (C) (D)
59. (A) (B) (C) (D)
60. (A) (B) (C) (D)

61. (A) (B) (C) (D)
62. (A) (B) (C) (D)
63. (A) (B) (C) (D)
64. (A) (B) (C) (D)
65. (A) (B) (C) (D)
66. (A) (B) (C) (D)
67. (A) (B) (C) (D)
68. (A) (B) (C) (D)
69. (A) (B) (C) (D)
70. (A) (B) (C) (D)

71. (A) (B) (C) (D)
72. (A) (B) (C) (D)
73. (A) (B) (C) (D)
74. (A) (B) (C) (D)
75. (A) (B) (C) (D)
76. (A) (B) (C) (D)
77. (A) (B) (C) (D)
78. (A) (B) (C) (D)
79. (A) (B) (C) (D)
80. (A) (B) (C) (D)

Practice Exam 2

41. A residential development is proposed along a rural highway that is adjacent to the open channel shown. The channel is 3 ft wide at the bottom and 3 ft deep. After development, the open channel is estimated to convey a peak flow of 102 ft³/sec during a 25-year storm event. While maintaining a maximum depth of 3 ft, the existing open channel swale has a capacity of 51 ft³/sec. The side slopes and freeboard will remain the same.

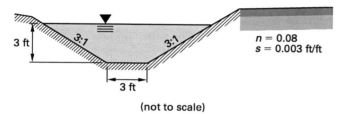

(not to scale)

Most nearly, how much wider does the swale bottom need to be to accommodate the estimated peak flow during the 25-year storm?

(A) 5.6 ft

(B) 8.8 ft

(C) 14 ft

(D) 17 ft

42. A 2 mi segment of a two-lane rural collector is designed to pass through pasture land. The design includes a new stormwater management facility. Pre-development and post-development hydrographs are shown. The stormwater management facility is designed to detain runoff with a peak discharge rate that matches the pre-development peak flow rate.

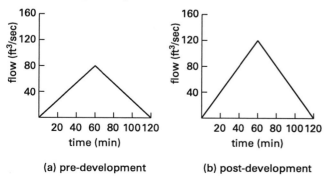

(a) pre-development   (b) post-development

Most nearly, what is the required storage of the proposed stormwater management facility?

(A) 1.1 ac-ft

(B) 2.2 ac-ft

(C) 3.3 ac-ft

(D) 4.3 ac-ft

43. A set of proposed safety improvement projects and their estimated effects on crash frequency and costs are shown. Using an incremental benefit-cost ratio analysis, determine the safety improvement project with the highest priority.

safety improvement project	estimated average annual reduction in crash frequency	present value of crash reduction	present value cost estimate
roundabout at Tampa Blvd. and Florida Ave.	120	$7,325,000	$3,120,000
traffic signal at Nebraska St. and University Dr.	100	$8,385,000	$4,250,000
traffic control at Racetrack Blvd.	50	$5,438,000	$1,245,000
road widening at Oregon Ave.	48	$8,324,000	$2,346,000

(A) roundabout at Tampa Blvd. and Florida Ave.

(B) traffic signal at Nebraska St. and University Dr.

(C) traffic control at Racetrack Blvd.

(D) road widening at Oregon Ave.

44. A roadway pavement with an asphalt top layer and a limestone base has full-depth cracks with evidence of base failure. Instead of replacing both the base and top layer, the roadway pavement is rehabilitated by using full-depth reclamation (FDR) and repaving with a 3 in top layer of hot mix asphalt. A cross section of the rehabilitated roadway pavement is shown. The FDR base layer was created by pulverizing the damaged asphalt layer, mixing it with the limestone base, and adding fly ash during the mix operations. The base resilient modulus with respect to the percentage of fly ash content of the FDR base is shown on the graph. The drainage coefficient for the FDR base and subgrade is 1.0. The required structural number for the roadway pavement is 4.5.

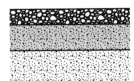

HMA, 3 in
strength coefficient = 0.4/in

FDR base, 10 in

subgrade, 12 in
strength coefficient = 0.1/in

(a) roadway section

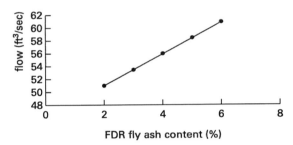

(b) average elastic resilient modulus versus
FDR fly ash content

According to AASHTO design guidelines, what percent of fly ash content needs to be added in FDR operations to obtain the required structural number?

(A)   2.0%

(B)   3.0%

(C)   4.5%

(D)   5.0%

45. A roadway is designed with a circular curve 376 ft long. The bearing from the point of curvature to the point of intersection is N15° 19′ 15″ E. The bearing from the point of intersection to the point of tangent is S12° 40′ 45″ E. Most nearly, which parameters correctly describe the circular curve?

(A) $I = 152°$, $R = 141.7$ ft, $T = 568.3$ ft

(B) $I = 28°$, $R = 8226.5$ ft, $T = 188.1$ ft

(C) $I = 28°$, $R = 769.40$ ft, $T = 191.8$ ft

(D) $I = 28°$, $R = 769.4$ ft, $T = 191.8$ ft

46. A signalized intersection has an actuated phase for through movements on the main road. The overall phase duration is 45 sec, the yellow change interval is 2.5 sec, the start-up lost time is 2 sec and the red clearance interval is 1 sec. The maximum queue size for the through movements during the peak hour is 18 vehicles. The queue discharge rate is 0.8 vehicles per second. According to the Transportation Research Board, what is most nearly the green extension time during the phase?

(A)   14.4 sec

(B)   17.0 sec

(C)   18.5 sec

(D)   22.5 sec

47. A vehicle travels at 45 mph during daylight on a roadway with wet pavement where the coefficient of skidding friction is 0.32. The driver sees an object blocking the roadway 325 ft ahead. The driver's brake reaction time is 1.5 sec. Which scenarios will allow the driver to avoid a collision with the object?

  I.   vehicle traveling on a +4% grade

  II.   vehicle traveling on a −4% grade

(A)   I only

(B)   II only

(C)   both I and II

(D)   neither I nor II

48. According to the MUTCD, how far in advance of a potential hazard should the following warning signs be placed?

  I.   "Signal Ahead" warning sign on a roadway with a posted speed limit of 55 mph

  II.   "Turn" warning sign with an advisory speed of 20 mph on a road with a posted speed limit of 40 mph

(A)   signal ahead = 325 ft; turn = 100 ft

(B)   signal ahead = 325 ft; turn = 670 ft

(C)   signal ahead = 990 ft; turn = 100 ft

(D)   signal ahead = 990 ft; turn = 670 ft

49. A roadway is designed with two curves as shown. Determine the station at the point of tangency for curve 2.

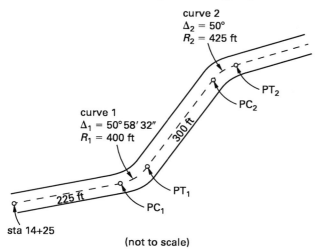

curve 2
$\Delta_2 = 50°$
$R_2 = 425$ ft

PT$_2$

PC$_2$

curve 1
$\Delta_1 = 50°58'32''$
$R_1 = 400$ ft

300 ft

PT$_1$

PC$_1$

225 ft

sta 14+25

(not to scale)

(A)   sta 20+06

(B)   sta 20+51

(C)   sta 25+95

(D)   sta 26+40

50. According to the AASHTO *Green Book*, which vehicle approaching an intersection needs the longest sight distance along the intersecting street?

I.   A combination truck on a minor road approaches an intersection and turns right from a stop. The design speed on the major road is 45 mph.

II.   A passenger car on a minor road approaches an intersection on a −6% grade, with no stop signs or signals. The design speed on the major road is 80 mph and 50 mph on the minor road.

III.   A passenger car on a minor road approaches an intersection, turning left from a stop. The design speed on the major road is 55 mph.

IV.   A single-unit truck on a minor road approaches an intersection and crosses through a 4-lane intersection from a stop. The design speed on the major road is 45 mph.

(A)   I

(B)   II

(C)   III

(D)   IV

51. A sag vertical curve has an incoming slope of −2% and an outgoing slope of +3%. The posted speed limit is 50 mph. Most nearly, what is the minimum length of the curve if the design criteria requires a 400 ft headlight sight distance?

(A)   350 ft

(B)   400 ft

(C)   450 ft

(D)   550 ft

52. A crest vertical curve connects grades of −2% and −4.75% as shown. At point Q, the station is sta 13+55 and the elevation is 210.6 ft. Most nearly, what is the elevation at point B on the curve at sta 17+00?

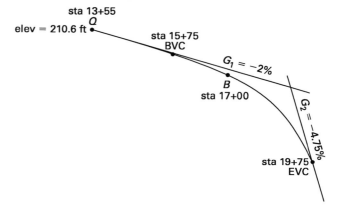

sta 13+55
Q
elev = 210.6 ft

sta 15+75
BVC

$G_1 = -2\%$

B
sta 17+00

$G_2 = -4.75\%$

sta 19+75
EVC

(A)   203.2 ft

(B)   204.2 ft

(C)   205.2 ft

(D)   206.2 ft

53. A roadway with a design speed of 50 mph has an intersection right-turn deceleration lane as shown. The right-turn lane storage length is 300 ft. An average driver has a brake reaction time of 2.5 sec. According to the AASHTO *Green Book*, what is most nearly the desirable taper distance?

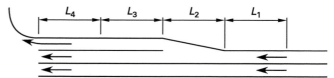

L_4 L_3 L_2 L_1

(A)   215 ft

(B)   300 ft

(C)   425 ft

(D)   610 ft

54. An engineering study is conducted to determine the traffic control signal warrant requirements for an intersection. The engineering study report indicates the following findings during an observed eight-hour period on an average day

   I. The total vehicles approaching on the major road is 6000 in both directions.

  II. The total vehicles approaching on the minor road is 1600 in both directions.

 III. The major road has three lanes in each direction.

 IV. The minor road has one lane in each direction.

  V. The directional split on the minor roadway is a consistent 60/40 during the study.

 VI. Alternative methods of remediation have been unsuccessful.

According to the MUTCD, which conditions of the eight-hour vehicular volume warrant are met for the 80% column?

(A) Condition A is satisfied.

(B) Condition B is satisfied.

(C) Both conditions A and B are satisfied.

(D) Neither conditions A or B are satisfied.

55. A contractor is analyzing two pavement striping machines for use over the next five years.

Machine 1 costs $18,000, requires $500 of annual maintenance, and has a salvage value of $7500.

Machine 2 costs $15,000, requires $700 of annual maintenance, and has a salvage value of $6000.

The annual interest rate is 8%. How much money can the contractor expect to save annually by selecting the machine with the lowest total annual cost?

(A) $150

(B) $195

(C) $240

(D) $295

56. Which of the following is NOT typically included in a traffic impact study as part of the traffic data?

(A) traffic counts

(B) traffic duration

(C) peak hours

(D) travel forecasting

57. According to the AASHTO *Guide for the Planning, Design, and Operation of Pedestrian Facilities*, how many times more likely is a pedestrian fatality when a vehicle hits a pedestrian at 40 mph versus 20 mph?

(A) 1.9

(B) 4.3

(C) 5.0

(D) 5.7

58. A horizontal curve is designed for a section of highway with a design speed of 50 mph. The point of tangency is at station 164+00, and the point of curvature is at station 150+00. If the central angle of the curve is 80°, what value of superelevation is required at this curve?

(A) 2.0%

(B) 2.6%

(C) 3.5%

(D) 4.2%

59. A crest vertical curve has an incoming grade of +3.0% and an outgoing grade of −1.0%. The beginning of the curve is at station 100+00 and an elevation of 920 ft. The curve is designed to have a minimum stopping sight distance of 520 ft, based on an object height of 2 ft. Most nearly, what is the curve's elevation at a distance of $\frac{1}{3}$ the curve length from the beginning of the curve?

(A) 924 ft

(B) 926 ft

(C) 931 ft

(D) 932 ft

60. A pedestrian walkway near a baseball stadium is designed to a level of service (LOS) C. The peak pedestrian flow rate is 1200 pedestrians in the peak 15 min period. According to the *Highway Capacity Manual* (HCM), most nearly, what is the minimum effective width of the walkway?

(A) 8 ft

(B) 10 ft

(C) 11 ft

(D) 15 ft

61. The data shown in the table was collected during a turning movement count for a northbound intersection approach during the p.m. peak hours. Most nearly, what is the peak hour factor, PHF, for the northbound approach?

time	peak period volumes (northbound)		
	left	through	right
3:30–3:45 p.m.	12	432	22
3:45–4:00 p.m.	15	451	26
4:00–4:15 p.m.	19	450	36
4:15–4:30 p.m.	25	475	48
4:30–4:45 p.m.	28	480	49
4:45–5:00 p.m.	22	50	56
5:00–5:15 p.m.	31	525	61
5:15–5:30 p.m.	35	536	75
5:30–5:45 p.m.	31	542	70
5:45–6:00 p.m.	25	515	55
6:00–6:15 p.m.	18	502	50

(A) 0.86

(B) 0.90

(C) 0.97

(D) 0.99

62. A traffic study conducted at a signalized intersection reports 274 vehicles making an eastbound left turn during the peak hour. The cycle length is 70 sec, and the effective green time for this movement is 12.5 sec. The saturated flow rate is 1800 veh/hr. The incremental delay has been calculated as 27 sec. The initial queue delay is zero, and the progression factor is 1. Most nearly, what is the level of service (LOS) for this movement?

(A) LOS A

(B) LOS B

(C) LOS C

(D) LOS D

63. According to the MUTCD, the components of a temporary traffic control zone, in the order in which

drivers encounter them on most publicly operated roadways, are

(A) advance warning, activity, transition, termination

(B) advance warning, upstream taper, buffer, downstream taper

(C) advance warning, transition, activity, termination

(D) advance warning, buffer, work space, buffer

64. According to the example data in Fig. 5-2(b) of the AASHTO *Roadside Design Guide* (RDG), which roadside embankment situation would most likely justify the installation of a guardrail at the top of an embankment slope?

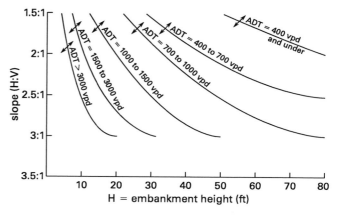

(a) AASHTO 5-2(b) design chart for embankment barriers

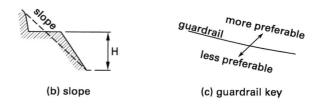

(b) slope

(c) guardrail key

(A) 2:1 slope, 25 ft embankment height, ADT = 750 vpd

(B) 2.5:1 slope, 50 ft embankment height, ADT = 500 vpd

(C) 1.5:1 slope, 15 ft embankment height, ADT = 1200 vpd

(D) 3.5:1 slope, 60 ft embankment height, ADT = 2550 vpd

65. A curve on a highway alignment has a higher than expected rate of crashes. The segment has an expected average annual daily traffic (AADT) of 5000 veh/day and a design speed of 50 mph. The cut and fill slopes at the study location are 10H:1V. According to AASHTO, what is most nearly the MINIMUM clear zone distance outside the curve for this location with a 1200 ft horizontal curve radius?

(A) 14 ft

(B) 18 ft

(C) 21 ft

(D) 30 ft

66. According to AASHTO, what is the preferred right-of-way distance for the following cross section of a high-speed, high-volume highway, in ft?

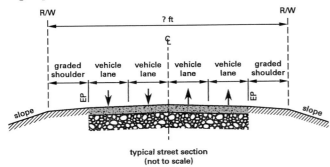

typical street section
(not to scale)

(A) 56 ft

(B) 64 ft

(C) 68 ft

(D) 72 ft

67. A bridge alignment is centered over the low point of a sag vertical curve that is 160 ft long. The roadway's sag curve has an incoming grade of −5% and an outgoing grade of 4.2%. The beginning of the vertical curve has an elevation of 513.15 ft at sta 10+00. The minimum vertical clearance between the center of the bridge structure and the roadway's low point is 20 ft. The minimum elevation of the bridge is most nearly

(A) 511 ft

(B) 519 ft

(C) 531 ft

(D) 539 ft

68. Through-movement vehicle data are collected for a single lane at a signalized intersection. Effective green time is 29 sec. Saturation headway is 2.2 sec/veh with a cycle length of 60 sec. Startup lost time is 2 sec. Clearance lost time is 1 sec. Amber time is 4 sec. According to

the *Highway Capacity Manual* (HCM), the through-movement capacity at the intersection is most nearly

(A) 709 vph

(B) 763 vph

(C) 791 vph

(D) 873 vph

69. A single-lane roadway entering a work zone with a speed limit of 45 mph requires a lane shift as shown, with an additional 2 ft for the placement of channelizing devices.

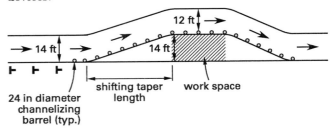

According to the *Manual on Uniform Traffic Control Devices* (MUTCD), what is the MINIMUM recommended shifting taper length?

(A) 100 ft

(B) 240 ft

(C) 300 ft

(D) 360 ft

70. According to AASHTO's *A Policy on Geometric Design of Highways and Streets*, which of the following design techniques generally reduce wrong-way movements at interchanges?

I. providing for all movements to and from the freeway

II. narrowing the arterial highway median opening

III. using conventional, easily recognizable interchange patterns

IV. merging on- and off-ramps that join a minor road

(A) I, II, and III

(B) I, II, and IV

(C) I, III, and IV

(D) II, III, and IV

71. A highway network is linked together at two nodes. The network has a free flow travel time of 47 min and a peak hour volume of 2200 veh/hr between nodes. Maximum peak hour capacity is 2400 veh/hr. Using the Bureau of Public Roads' function and substituting base values of $a = 0.15$ and $b = 4.0$, the total travel time between the two nodes is most nearly

(A) 42 min

(B) 47 min

(C) 52 min

(D) 57 min

72. The preferred mode of transportation for 10,000 commuters between two zones is analyzed. Using a modal split model, the value of a mode of transportation to the commuter is

$$v_m = -0.006t_m - 0.004c_m - 0.003\,W_m$$

v_m is the value of mode m to the commuter. t_m is the commute time for mode m. c_m is the cost to commute by mode m. W_m is the wait time for mode m. Commute times and costs for each mode of transportation are given in the table.

travel mode, m	t_m (min)	c_m ($)	W_m (min)
automobile	45	8	0
carpool	55	4	10
bus	65	2	20

The number of commuters expected to commute by bus is most nearly

(A) 3070

(B) 3335

(C) 3590

(D) 4580

73. The number of southbound left turns at an intersection is given in the table.

signal cycle	left turns (southbound)
1	6
2	2
3	7
4	8
5	9
6	11
7	6
8	5

The standard error for the data set is most nearly

(A) 0.88

(B) 0.90

(C) 0.96

(D) 0.99

74. An intersection with a has 264 recorded crashes per year. 25% of these accidents are rear-end type. After implementing a countermeasure, the number of rear-end crashes is reduced by 45 crashes per year. The rear-end crash reduction factor for the countermeasure is most nearly

(A) 0.32

(B) 0.68

(C) 0.83

(D) 0.92

75. An all-way stop controlled T-intersection has the peak hour volumes and control delays given in the table.

approach	left	through	right	control delay (sec)
eastbound	72	400	–	12.0
southbound	115	–	36	13.5
westbound	–	365	125	14.0

The peak hour factor for the intersection is 0.85. What is level of service (LOS) for the intersection?

(A) LOS A

(B) LOS B

(C) LOS C

(D) LOS D

76. A pedestrian overpass is designed to cross a major highway. The difference in elevation between the surface of the roadway and the top of the overpass is 17.5 ft. In addition to stairs, the overpass has accessible pedestrian ramps with landings that have no longitudinal slope. According to the AASHTO *Guide for Planning and Design of Pedestrian Facilities*, the pedestrian ramp requires how many level landings to be ADA compliant? (Include the top landing.)

(A) 4

(B) 5

(C) 6

(D) 7

77. A flexible pavement section that is 8 in thick with a terminal serviceability index of 2.5 is subjected to loading from truck A and truck B. The pavement has a structural number of 5.

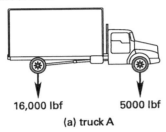

(a) truck A

16,000 lbf   5000 lbf

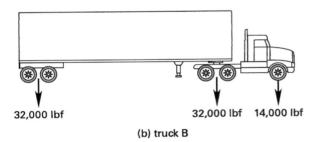

(b) truck B

32,000 lbf   32,000 lbf   14,000 lbf

According to AASHTO, the pavement damage from a single pass of truck B is most nearly equivalent to how many passes of truck A?

(A) 2.50

(B) 3.30

(C) 3.70

(D) 5.10

78. The data from a standard soil compaction test are shown.

water content (%)	dry unit weight (lbf/ft³)
6	102
10	107
13	109
16	106
19	102

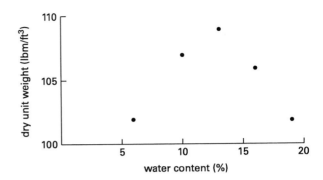

Most nearly, what is the optimum water content?

(A) 6%

(B) 10%

(C) 13%

(D) 19%

79. During a signal warrant analysis, a weekday 14 hr traffic count (6:00 am to 8:00 pm) with turning movements was collected for an existing stop-controlled intersection near a local factory. Stopped time delay data was also collected. All roadways at the intersection have one-lane approaches. Data for the peak a.m. and peak p.m. hours are shown in the table.

	EB (veh/hr)	SB (veh/hr)	NB (veh/hr)	WB (veh/hr)
a.m. stopped delay	0.25	1.20	3.00	0.40
p.m. stopped delay	1.50	3.50	1.40	0.60
approach volume	EB (veh)	SB (veh)	NB (veh)	WB (veh)
6:00 a.m.	15	45	105	25
6:15 a.m.	30	60	108	20
6:30 a.m.	10	62	90	15
6:45 a.m.	20	70	110	5
approach volume	EB (veh)	SB (veh)	NB (veh)	WB (veh)
4:00 p.m.	59	82	71	42
4:15 p.m.	62	96	82	39
4:30 p.m.	67	120	80	51
4:45 p.m.	64	138	85	52

For which time period is the MUTCD peak hour vehicular warrant satisfied?

(A) a.m. peak hour

(B) p.m. peak hour

(C) both a.m. and p.m. peak hour

(D) neither a.m. nor p.m. peak hour

80. Different asphalt mix designs for a 2 in thick asphaltic surface course have been submitted by the contractor to the engineer for review as shown.

mix design	percent asphalt, P_b	percent aggregate, P_s	specific gravity of asphalt, G_b	specific gravity of aggregate, G_{se}
A	5	95	2.3	2.5
B	6	94	2.4	2.5
C	7	93	2.2	2.5
D	8	92	2.4	2.6

The design plan estimates a spread rate for the structural layer asphalt of 220 lbm/yd². The spread rate equation of asphalt mix with thickness, t, in inches is shown.

$$\text{spread rate} \left(\frac{\text{lbm}}{\text{yd}^2} \right) = t G_{\text{mm}} \left(43.3 \ \frac{\text{lbm}}{\text{in-yd}^2} \right)$$

Which submitted mix design will require a higher spread rate than the spread rate estimated in the plans?

(A) mix design A

(B) mix design B

(C) mix design C

(D) mix design D

STOP!

DO NOT CONTINUE!

This concludes the Afternoon Session of the examination. If you finish early, check your work and make sure that you have followed all instructions. After checking your answers, you may turn in your examination booklet and answer sheet and leave the examination room. Once you leave, you will not be permitted to return to work or change your answers.

Practice Exam Answer Key

Practice Exam 1 Answer Key

1. B	11. C	21. A	31. B
2. C	12. B	22. A	32. C
3. C	13. C	23. C	33. A
4. B	14. D	24. D	34. B
5. C	15. D	25. D	35. C
6. D	16. A	26. B	36. D
7. B	17. A	27. D	37. A
8. C	18. B	28. C	38. D
9. D	19. A	29. C	39. B
10. B	20. B	30. C	40. C

Practice Exam 2 Answer Key

41. B	51. C	61. C	71. C
42. A	52. B	62. D	72. A
43. D	53. A	63. C	73. C
44. D	54. C	64. C	74. B
45. A	55. D	65. C	75. B
46. B	56. B	66. D	76. D
47. A	57. D	67. C	77. B
48. A	58. B	68. C	78. C
49. D	59. A	69. D	79. D
50. A	60. A	70. A	80. D

Solutions
Practice Exam 1

1. Use the 10-year storm frequency curve to find the rainfall intensity for the existing time of concentration.

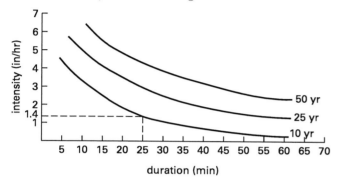

The total area of the existing contributory basin is

$$A_{\text{basin}} = 5.2 \text{ ac} + 10.3 \text{ ac} + 3.1 \text{ ac}$$
$$= 18.6 \text{ ac}$$

The peak runoff rate under existing conditions is

$$Q = C_e i A$$
$$= (0.54)\left(1.4 \ \frac{\text{in}}{\text{hr}}\right)(18.6 \text{ ac})$$
$$= 14.1 \text{ ft}^3/\text{sec}$$

The area of added pavement is

$$A_{\text{added pavement}} = 2 W_{\text{lane}} L_{\text{lane}}$$
$$= \frac{(2)(12 \text{ ft})(1 \text{ mi})\left(5280 \ \dfrac{\text{ft}}{\text{mi}}\right)}{\left(43{,}560 \ \dfrac{\text{ft}^2}{\text{ac}}\right)}$$
$$= 2.9 \text{ ac}$$

The total area for highway land use under proposed conditions is

$$A_{\text{pavement}} = A_{\text{existing pavement}} + A_{\text{added pavement}}$$
$$= 3.1 \text{ ac} + 2.9 \text{ ac}$$
$$= 6.0 \text{ ac}$$

The total area for meadow land use under proposed conditions is

$$A_{\text{meadow}} = A_{\text{existing meadow}} + A_{\text{added pavement}}$$
$$= 5.2 \text{ ac} - 2.9 \text{ ac}$$
$$= 2.3 \text{ ac}$$

The composite runoff coefficient under proposed conditions is

$$C_r = C_{\text{meadow}}\left(\frac{A_{\text{meadow}}}{A_{\text{basin}}}\right) + C_{\text{residential}}\left(\frac{A_{\text{residential}}}{A_{\text{basin}}}\right)$$
$$+ C_{\text{highway}}\left(\frac{A_{\text{pavement}}}{A_{\text{basin}}}\right)$$
$$= (0.40)\left(\frac{2.3 \text{ ac}}{18.6 \text{ ac}}\right) + (0.50)\left(\frac{10.3 \text{ ac}}{18.6 \text{ ac}}\right)$$
$$+ (0.90)\left(\frac{6.0 \text{ ac}}{18.6 \text{ ac}}\right)$$
$$= 0.62$$

The proposed time of concentration is

$$t_p = 0.80 t_e$$
$$= (0.80)(25 \text{ min})$$
$$= 20 \text{ min}$$

Find the rainfall intensity under proposed conditions.

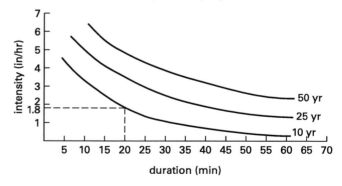

The peak flow under proposed conditions is

$$Q = C_r i A$$
$$= (0.62)\left(1.8 \ \frac{\text{in}}{\text{hr}}\right)(18.6 \ \text{ac})$$
$$= 20.8 \ \text{ft}^3/\text{sec}$$

The increase in peak runoff due to roadway widening is

$$\Delta Q = 20.8 \ \frac{\text{ft}^3}{\text{sec}} - 14.1 \ \frac{\text{ft}^3}{\text{sec}}$$
$$= 6.70 \ \text{ft}^3/\text{sec} \quad (7 \ \text{ft}^3/\text{sec})$$

The answer is (B).

2. The amount of equivalent crossing lanes is

$$\text{equivalent crossing lanes} = \text{lanes to cross} + \frac{\text{width of median}}{12 \ \frac{\text{ft}}{\text{ln}}}$$
$$= 1 \ \text{ln} + \frac{24 \ \text{ft}}{12 \ \frac{\text{ft}}{\text{ln}}}$$
$$= 3 \ \text{ln}$$

From Table 9-5 of the AASHTO *Green Book*, the time gap for a single-unit truck is 9.5 sec plus 0.7 sec for each equivalent crossing lane. The adjusted time gap is

$$t_{g,\text{adjusted}} = t_g + \left(0.7 \ \frac{\text{sec}}{\text{ln}}\right)(\text{equivalent crossing lanes})$$
$$= 9.5 \ \text{sec} + \left(0.7 \ \frac{\text{sec}}{\text{ln}}\right)(3 \ \text{ln})$$
$$= 11.6 \ \text{sec}$$

Using AASHTO *Green Book* Eq. 9-1, the minimum required intersection sight distance for the turn is

$$\text{ISD} = 1.47 V_{\text{major}} t_{g,\text{adjusted}}$$
$$= (1.47)\left(45 \ \frac{\text{mi}}{\text{hr}}\right)(11.6 \ \text{sec})$$
$$= 767 \ \text{ft} \quad (770 \ \text{ft})$$

(Units are not consistent in this empirically derived formula.)

The answer is (C).

3. Using the average end area method, the cut volume between sta 10+00 and sta 11+00 is

$$V_{\text{cut}} = (\text{sta}_{\text{end}} - \text{sta}_{\text{begin}})\left(\frac{A_{\text{begin cut}} + A_{\text{end cut}}}{2}\right)$$
$$= \left(\frac{(11 \ \text{sta} - 10 \ \text{sta})\left(100 \ \frac{\text{ft}}{\text{sta}}\right)}{3 \ \frac{\text{ft}}{\text{yd}}}\right)\left(\frac{10 \ \text{yd}^2 + 1 \ \text{yd}^2}{2}\right)$$
$$= 184 \ \text{yd}^3$$

The fill volume between sta 10+00 and sta 11+00 is

$$V_{\text{fill}} = (\text{sta}_{\text{end}} - \text{sta}_{\text{begin}})\left(\frac{A_{\text{begin fill}} + A_{\text{end fill}}}{2}\right)$$
$$= \left(\frac{\left(11 \ \text{sta} - 10 \ \text{sta}\right)\left(100 \ \frac{\text{ft}}{\text{sta}}\right)}{3 \ \frac{\text{ft}}{\text{yd}}}\right)\left(\frac{20 \ \text{yd}^2 + 30 \ \text{yd}^2}{2}\right)$$
$$= 833 \ \text{yd}^3$$

The net earthwork between sta 10+00 and sta 11+00 is

$$V_{\text{net}} = V_{\text{fill}} - V_{\text{cut}}$$
$$= 833 \ \text{yd}^3 - 184 \ \text{yd}^3$$
$$= 649 \ \text{yd}^3$$

Repeat these steps for the next stations to determine the cut, fill, and net volumes.

station	cut area (yd^2)	fill area (yd^2)	cut volume (yd^3)	fill volume (yd^3)	net volume (yd^3)
10+00	10	20	0	0	0
11+00	1	30	183.3	833.3	650.0
12+00	5	10	100.0	666.7	566.7
13+00	6	6	183.3	266.7	83.3
14+00	10	5	266.7	183.3	-83.3
15+00	12	8	366.7	216.7	-150.0
16+00	10	7	366.7	250.0	-116.7
17+00	20	2	500.0	150.0	-350.0
18+00	20	2	666.7	66.7	-600.0
19+00	30	2	833.3	66.7	-766.7

Find the cumulative earthwork volume between all stations. The cumulative earthwork volume at sta 11+00 is

$$V_{\text{cumulative sta 11+00}} = V_{\text{cumulative sta 10+00}} + V_{\text{net sta 11+00}}$$
$$= 0 \text{ yd}^3 + 650 \text{ yd}^3$$
$$= 650 \text{ yd}^3$$

Continue calculating the cumulative earthwork volume (CEV) for each station.

station	cut area (yd^2)	fill area (yd^2)	cut volume (yd^3)	fill volume (yd^3)	net volume (yd^3)	CEV (yd^3)
10+00	10	20	0	0	0	0
11+00	1	30	183.3	833.3	650.0	650.0
12+00	5	10	100.0	666.7	566.7	1216.7
13+00	6	6	183.3	266.7	83.3	1300.0
14+00	10	5	266.7	183.3	-83.3	1216.7
15+00	12	8	366.7	216.7	-150.0	1066.7
16+00	10	7	366.7	250.0	-116.7	950.0
17+00	20	2	500.0	150.0	-350.0	600
18+00	20	2	666.7	66.7	-600.0	0
19+00	30	2	833.3	66.7	-766.7	-766.7

The first balanced station occurs when the cumulative volume first equals zero, which occurs at sta 18+00.

The answer is (C).

4. Using the AASHTO layer-thickness equation, the structural number of the pavement section is

$$\text{SN} = D_1 a_1 + D_2 a_2 m_2 + D_3 a_3 m_3$$
$$= (5 \text{ in})\left(0.44 \frac{1}{\text{in}}\right) + (8 \text{ in})\left(0.12 \frac{1}{\text{in}}\right)(1.0)$$
$$\quad + (12 \text{ in})\left(0.07 \frac{1}{\text{in}}\right)(1.0)$$
$$= 4.0$$

Use Fig. 3-1 in Sec. II of the AASHTO *Guide for Design of Pavement Structures* to find the ESAL. With a structural number of 4.0, a design serviceability loss of 1, a reliability of 99% with a standard deviation of 0.2, and a resilient modulus of 15 kips/in^2 for the roadbed, the design ESAL is 6.0×10^6.

The answer is (B).

5. Rearrange the AASHTO layer-thickness equation to find the minimum asphalt layer thickness.

$$\text{SN} = D_1 a_1 + D_2 a_2 m_2 + D_3 a_3 m_3$$
$$D_1 = \frac{\text{SN} - D_2 a_2 m_2 - D_3 a_3 m_3}{a_1}$$
$$= \frac{4.4 - (10 \text{ in})\left(0.2 \frac{1}{\text{in}}\right)(1.0) - (12 \text{ in})\left(0.1 \frac{1}{\text{in}}\right)(1.0)}{0.4 \frac{1}{\text{in}}}$$
$$= 3.0 \text{ in}$$

The answer is (C).

6. Using the MUTCD Fig. 9C-8, the taper length requirement for the obstruction pavement markings is

$$L = (W + 1 \text{ ft})S$$
$$= (1 \text{ ft} + 1 \text{ ft})(20 \text{ mph})$$
$$= 40 \text{ ft}$$

The answer is (D).

7. Using the inlet efficiencies chart, the intercepted flow at inlet S-1 is

$$Q_{\text{intercepted,S-1}} = 2.75 \frac{\text{ft}^3}{\text{sec}}$$

The flow bypass at inlet 1 is

$$Q_{\text{bypass,S-1}} = Q_{\text{total,S-1}} - Q_{\text{intercepted,S-1}}$$
$$= 3.0 \frac{\text{ft}^3}{\text{sec}} - 2.75 \frac{\text{ft}^3}{\text{sec}}$$
$$= 0.25 \text{ ft}^3/\text{sec}$$

The total peak flow at inlet S-2 is

$$Q_{\text{total,S-2}} = Q_{\text{subbasin,S-2}} + Q_{\text{bypass,S-1}}$$
$$= 3.50 \frac{\text{ft}^3}{\text{sec}} + 0.25 \frac{\text{ft}^3}{\text{sec}}$$
$$= 3.75 \text{ ft}^3/\text{sec}$$

Using the inlet efficiencies chart, the intercepted flow at inlet S-2 is

$$Q_{\text{intercepted,S-2}} = 1.40 \frac{\text{ft}^3}{\text{sec}}$$

The flow bypass at inlet S-2 is

$$Q_{\text{bypass,S-2}} = Q_{\text{total,S-2}} - Q_{\text{intercepted,S-2}}$$
$$= 3.75 \ \frac{\text{ft}^3}{\text{sec}} - 1.40 \ \frac{\text{ft}^3}{\text{sec}}$$
$$= 2.35 \ \text{ft}^3/\text{sec}$$

The total peak flow at inlet S-3 is

$$Q_{\text{total S-3}} = Q_{\text{subbasin S-3}} + Q_{\text{bypass S-2}}$$
$$= 2.80 \ \frac{\text{ft}^3}{\text{sec}} + 2.35 \ \frac{\text{ft}^3}{\text{sec}}$$
$$= 5.15 \ \text{ft}^3/\text{sec}$$

Using the integrated Manning's equation, the maximum gutter spread between sta 14+50 and sta 18+00 is

$$Q = \left(\frac{0.56}{n}\right) S_x^{5/3} S_L^{\frac{1}{2}} T^{8/3}$$

$$T = \left(\frac{nQ}{0.56 S_x^{5/3} S_L^{1/2}}\right)^{3/8}$$

$$= \left(\frac{(0.012)\left(5.15 \ \dfrac{\text{ft}^3}{\text{sec}}\right)}{(0.56)\left(0.02 \ \dfrac{\text{ft}}{\text{ft}}\right)^{5/3}\left(0.015 \ \dfrac{\text{ft}}{\text{ft}}\right)^{1/2}}\right)^{3/8}$$

$$= 11.1 \ \text{ft} \quad (11 \ \text{ft})$$

The answer is (B).

8. From the graph, the maximum dry density is at the apex of the curve, which is 120 lbm/ft^3.

The dry density of sample 1 is

$$\rho_d = \frac{\rho}{1+w}$$
$$= \frac{120 \ \dfrac{\text{lbm}}{\text{ft}^3}}{1+0.06}$$
$$= 113.2 \ \text{lbm/ft}^3$$

The percent compaction of sample 1 is

$$\frac{113.2 \ \dfrac{\text{lbm}}{\text{ft}^3}}{120.0 \ \dfrac{\text{lbm}}{\text{ft}^3}} \times 100\% = 94\%$$

Find the percent compaction for each sample.

sample	density (lbm/ft^3)	percent moisture content	dry density (lbm/ft^3)	percent compaction
1	120	6	113.2	94
2	120	17	102.6	85
3	125	5	119.0	99
4	125	13	110.6	92

Sample 3 is above 98% compaction.

The answer is (C).

9. The number of trips per household is

$$T = 0.78 + 1.6P + 2.4A$$
$$= 0.78 + (1.6)\left(3 \ \frac{\text{persons}}{\text{house}}\right) + (2.4)\left(2.2 \ \frac{\text{veh}}{\text{house}}\right)$$
$$= 10.9 \ \text{trips per day-house}$$

The average number of TAZ trips per day is

$$T_{\text{zone}} = (425 \ \text{houses})\left(10.9 \ \frac{\text{trips}}{\text{day-house}}\right)$$
$$= 4616 \ \text{trips/day} \quad (4600 \ \text{trips/day})$$

The answer is (D).

10. From HCM Chap. 15, Class II highways are mostly collectors and local roads that can serve as scenic or recreational routes. Since Class II highways are mostly collectors, and since motorists on local roads do not expect to travel at high speeds, Class II highways often serve relatively short trips.

Class III highways serve moderately developed areas. Class I highways serve as primary connectors of major traffic generators, daily commuter routes, major city routes, or major links in state or national highway networks.

The answer is (B).

11. From AASHTO *Green Book* Table 9-2, a single-lane roundabout has a recommended maximum entry design speed of 20 to 25 mph, a typical inscribed circle diameter of 90 to 150 ft, and a typical daily volume of 20,000 or fewer vehicles per day. The roundabout is single-lane.

Based on the description of the roundabout, the inscribed diameter is too large for a mini-roundabout and too small for a multilane.

The three basic types of roundabout are mini, single-lane, and multilane. A roundabout in an urban area may be any one of these three.

The answer is (C).

12. Rearrange the equation for braking/skidding distance to find the speed of the vehicle when the vehicle enters the grass. Use the collision speed for the final velocity. Since the road is level, the decimal grade is zero.

$$s_{grass} = \frac{v_{grass}^2 - v_{collision}^2}{30(f + G)}$$

$$v_{grass} = \sqrt{s_{grass}\left(30(f + G)\right) + v_{collision}^2}$$

$$= \sqrt{(60 \text{ ft})\left((30)(0.5 + 0)\right) + \left(25 \, \frac{\text{mi}}{\text{hr}}\right)^2}$$

$$= 39.1 \text{ mi/hr}$$

Find the speed of the vehicle when braking began. Use the speed of the vehicle when it entered the grass for the final velocity.

$$v_{asphalt} = \sqrt{s_{asphalt}\left(30(f_{asphalt} + G)\right) + v_{grass}^2}$$

$$= \sqrt{(130 \text{ ft})\left((30)(0.8 + 0)\right) + \left(39.1 \, \frac{\text{mi}}{\text{hr}}\right)^2}$$

$$= 68.2 \text{ mph} \quad (68 \text{ mph})$$

The answer is (B).

13. If the sight stopping distance, S, is greater than the length of the vertical curve, L, the equation for vertical curve length is

$$L = 2S - \frac{200\left(\sqrt{h_1} + \sqrt{h_2}\right)^2}{A}$$

Rearranging, the sight stopping distance is

$$S = \frac{L}{2} + \frac{200\left(\sqrt{h_1} + \sqrt{h_2}\right)^2}{2A}$$

$$= \frac{900 \text{ ft}}{2} + \frac{(200)\left(\sqrt{3.5 \text{ ft}} + \sqrt{0.50 \text{ ft}}\right)^2}{(2)\left(3.25 \, \frac{\text{ft}}{\text{ft}}\right)}$$

$$= 655 \text{ ft}$$

However, $S < L$, so the equation for vertical curve length when $S > L$ is not applicable.

Use the equation for vertical curve length when $S < L$.

$$L = \frac{AS^2}{200\left(\sqrt{h_1} + \sqrt{h_2}\right)^2}$$

Rearranging, the sight stopping distance is

$$S = \left(\frac{200L\left(\sqrt{h_1} + \sqrt{h_2}\right)^2}{A}\right)^{0.5}$$

$$= \left(\frac{(200)(900 \text{ ft})\left(\sqrt{3.5 \text{ ft}} + \sqrt{0.50 \text{ ft}}\right)^2}{3.25 \, \frac{\text{ft}}{\text{ft}}}\right)^{0.5}$$

$$= 607 \text{ ft} \quad (610 \text{ ft})$$

Since $S < L$, the condition is satisfied.

The answer is (C).

14. From AASHTO *Green Book* Table 9-7, the time gap for a combination truck on a minor road to make a right turn onto a major road at a stop-controlled intersection is 10.5 sec.

From AASHTO Table 9-4, the approach grade of the minor road is in excess of a 3% upgrade, so 0.1 sec is added for each percent. The adjusted time gap is

$$t_{g,adjusted} = t_g + (4\%)\left(0.1 \, \frac{\text{sec}}{\%}\right)$$

$$= 10.5 \text{ sec} + (4\%)\left(0.1 \, \frac{\text{sec}}{\%}\right)$$

$$= 10.9 \text{ sec}$$

The minimum sight distance along the major roadway for the truck to safely turn right is

$$\text{ISD} = 1.47\, V_{major}\, t_{g,adjusted}$$

$$= (1.47)\left(45 \, \frac{\text{mi}}{\text{hr}}\right)(10.9 \text{ sec})$$

$$= 721 \text{ ft}$$

The answer is (D).

15. From HCM Exh. 22-16, the percentage of total traffic volume in the eastbound leftmost lane is

$$\% V_{L1} = \frac{1}{3} - 0.245\left(\frac{\nu_F}{\nu_E + \nu_F + \nu_I}\right) + 0.465\left(\frac{\nu_E}{\nu_E + \nu_F + \nu_I}\right)$$
$$+ 0.465\left(\frac{\nu_E}{\nu_E + \nu_F + \nu_I}\right)$$

$$= \left(\frac{1}{3} - (0.245)\left(\frac{810 \; \frac{\text{veh}}{\text{hr}}}{800 \; \frac{\text{veh}}{\text{hr}} + 810 \; \frac{\text{veh}}{\text{hr}} + 750 \; \frac{\text{veh}}{\text{hr}}}\right)\right.$$
$$+ (0.465)\left(\frac{800 \; \frac{\text{veh}}{\text{hr}}}{800 \; \frac{\text{veh}}{\text{hr}} + 810 \; \frac{\text{veh}}{\text{hr}} + 750 \; \frac{\text{veh}}{\text{hr}}}\right)\right)$$
$$\times (100\%)$$
$$= 41\%$$

The answer is (D).

16. Refer to Table 3-11b from AASHTO's *A Policy on Geometric Design of Highways and Streets* (GDHS). When the maximum superelevation rate is 10%, the curve radius 2200 ft, and the design speed 60 mph, the full superelevation rate is $e = 7\%$.

From Table 3-17b , the runoff for a roadway with one rotated lane, a 7% rate of superelevation, and a design speed of 60 mph is 184 ft.

Using GDHS Eq. 3-24, the runout is

$$L_t = \left(\frac{e_{NC}}{2e_d}\right)L_r$$
$$= \left(\frac{2 \; \frac{\text{ft}}{\text{ft}}}{7 \; \frac{\text{ft}}{\text{ft}}}\right)(184 \text{ ft})$$
$$= 52.5 \text{ ft} \quad (53 \text{ ft})$$

The answer is (A).

17. The MUTCD, Sec. 4C.10 Warrant 9, Intersection Near a Grade Crossing, Paragraph 9, lists three standards that must be met if Warrant 9 is used to justify the installation of a traffic control signal. These include flashing-light signals, preemption, and actuation. Although Paragraph 10 recommends that automatic gates be installed at grade crossings if Warrant 9 is invoked, this is only a recommendation and not a standard.

The answer is (A).

18. The present worth for the $1000 cost component is

$$P_{P/A} = A\frac{(1+i)^n - 1}{i(1+i)^n}$$
$$= (\$1000)\left(\frac{(1+0.07)^{10} - 1}{(0.07)(1+0.07)^{10}}\right)$$
$$= \$7024$$

The present worth for the $250 gradient cost component is

$$P_G = G\left(\frac{(1+i)^n - 1}{i^2(1+i)^n} - \frac{n}{i(1+i)^n}\right)$$
$$= (\$250)\left(\frac{(1+0.07)^{10} - 1}{(0.07)^2(1+0.07)^{10}} - \frac{10}{(0.07)(1+0.07)^{10}}\right)$$
$$= \$6929$$

The present worth of the maintenance costs over the lifetime of the new pavement is

$$P = P_{P/A} + P_G$$
$$= \$7024 + \$6929$$
$$= \$13,953 \quad (\$14,000)$$

The answer is (B).

19. Treating utility poles along a corridor to minimize the likelihood of utility pole crashes is listed as objective C in RDG Table 4-1. Placing utilities underground, relocating poles farther from the roadway, and decreasing the number of poles along a corridor are all strategies for accomplishing objective C. Using breakaway poles is a strategy suggested only for objective A.

The answer is (A).

20. The lateral extent of the area of concern, L_A, is the distance from the edge of the traveled way to the far side of the fixed object.

$$L_A = 15 \text{ ft} + 3 \text{ ft}$$
$$= 18 \text{ ft}$$

From RDG Table 5-10(b), the runout length is $L_R = 250$ ft. From Fig. 5-39, the given length-of-need, X, is 105 ft.

Using RDG Eq. 5.3, the lateral offset of the guardrail is

$$Y = L_A - \left(\frac{L_A}{L_R}\right)X$$
$$= 18 \text{ ft} - \frac{(18 \text{ ft})(105 \text{ ft})}{250 \text{ ft}}$$
$$= 10.4 \text{ ft} \quad (10 \text{ ft})$$

The answer is (B).

21. The width for the proposed roadway is

$$W_{\text{proposed}} = 2(W_{\text{clear zone}} + W_{\text{bike lane}} + W_{\text{lane 1}} + W_{\text{lane 2}})$$
$$+ W_{\text{median}}$$
$$= (2)(12 \text{ ft} + 8 \text{ ft} + 12 \text{ ft} + 14 \text{ ft}) + 18 \text{ ft}$$
$$= 110 \text{ ft}$$

The additional right-of-way required on each side of the street to fit the design parameters is

$$\Delta W_{\text{proposed}} = \frac{W_{\text{proposed}} - W_{\text{existing}}}{2}$$
$$= \frac{110 \text{ ft} - 88 \text{ ft}}{2}$$
$$= 11 \text{ ft}$$

Lane width is measured from center of lane stripe to center of lane stripe, thus the width of the striping is not relevant. The curb height is not needed because curb height relates to vertical change, and the problem relates only to horizontal measurements. Bike lane width is measured from center of the bike lane stripe to the curb line. Gutter width is always included as part of the bike lane dimension. Sidewalk width is part of the parkway width.

The answer is (A).

22. Use HCM Eq. 17-6. Start-up lost time for signalized intersections is 2.0 sec. The segment running time is

$$t_R = \left(\frac{6.0 - l_1}{0.0025L}\right)f_x + \left(\frac{3600L}{5280S_f}\right)f_v + \sum_{i=1}^{N_{\text{ap}}} d_{\text{ap},i} + d_{\text{other}}$$

$$= \left(\frac{6.0 \text{ sec} - 2.0 \text{ sec}}{(0.0025)(0.5 \text{ mi})\left(5280 \frac{\text{ft}}{\text{mi}}\right)}\right)(1.0)$$

$$+ \left(\frac{\left(3600 \frac{\text{sec}}{\text{hr}}\right)(0.5 \text{ mi})\left(5280 \frac{\text{ft}}{\text{mi}}\right)}{\left(5280 \frac{\text{ft}}{\text{mi}}\right)\left(40 \frac{\text{mi}}{\text{hr}}\right)}\right)(1.01)$$

$$+ (6)(22 \text{ sec}) + (0 \text{ sec})$$
$$= 178.06 \text{ sec} \quad (180 \text{ sec})$$

The answer is (A).

23. The ramp length is

$$L_{\text{ramp}} = \frac{\Delta H}{\text{slope}}$$
$$= \frac{8 \text{ in}}{\left(\frac{1 \text{ in}}{12 \text{ in}}\right)\left(12 \frac{\text{in}}{\text{ft}}\right)}$$
$$= 8 \text{ ft}$$

The minimum horizontal distance from MOC perpendicular to BOW is

$$L_{\text{total}} = L_{\text{ramp}} + L_{\text{landing}}$$
$$= 8 \text{ ft} + 4 \text{ ft}$$
$$= 12 \text{ ft}$$

The answer is (C).

24. Rearrange HCM Eq. 21-1 to find the conflicting flow rate.

$$c_{e,\text{pce}} = 1130e^{(-1.0 \times 10^{-3})v_{c,\text{pce}}}$$

$$v_{c,\text{pce}} = \frac{\ln\dfrac{c_{e,\text{pce}}}{1130}}{-1.0 \times 10^{-3}}$$

$$= \frac{\ln\dfrac{460 \text{ pcph}}{1130}}{-1.0 \times 10^{-3}}$$
$$= 898.7 \text{ pcph}$$

Using the HCM Eq. 21-3, the new lane capacity after adding the second circulating lane is

$$c_{p,\text{pce}} = 1130e^{(-0.7x10^{-3})v_{c,\text{pce}}}$$
$$= 1130e^{(-0.7x10^{-3})(898.7 \text{ pcph})}$$
$$= 602.4 \text{ pcph} \quad (600 \text{ pcph})$$

The answer is (C).

25. Using the equation for tangent length, the interior angle is

$$T = R \tan\frac{I}{2}$$
$$I = 2\arctan\frac{T}{R}$$
$$= 2\arctan\frac{600 \text{ ft}}{5000 \text{ ft}}$$
$$= 13.7°$$

The length of the curve is

$$L = \frac{2\pi R I}{360°}$$
$$= \frac{(2\pi)(5000 \text{ ft})(13.7°)}{360°}$$
$$= 1196 \text{ ft}$$

The station at the point of curvature is

$$\text{sta}_{\text{PC}} = \text{sta}_{\text{PI}} - T$$
$$= \text{sta } 204{+}00 - 600 \text{ ft}$$
$$= \text{sta } 198{+}00$$

The station at the point of tangency is

$$\text{sta}_{\text{PT}} = \text{sta}_{\text{PC}} + L$$
$$= \text{sta } 198{+}00 + 1196 \text{ ft}$$
$$= \text{sta } 209{+}96$$

The answer is (D).

26. Find the horizontal sightline offset (HSO) by calculating the distance between the edge of the billboard and the center of the nearest lane.

$$\text{HSO} = 77 \text{ ft} - \frac{12 \text{ ft}}{2} - \frac{50 \text{ ft}}{2} = 46 \text{ ft}$$

The curve radius is

$$R = \frac{5729.578 \text{ deg-ft}}{D}$$
$$= \frac{5729.578 \text{ deg-ft}}{9°}$$
$$= 636.62 \text{ ft}$$

Rearrange the HSO equation to find the minimum sight distance along the curve.

$$\text{HSO} = R\left(1 - \cos\frac{RS}{28.65}\right)$$
$$S = \frac{R}{28.65}\arccos\frac{R - \text{HSO}}{R}$$
$$= \frac{636.62 \text{ ft}}{28.65}\arccos\frac{636.62 \text{ ft} - 46 \text{ ft}}{636.62 \text{ ft}}$$
$$= 486.9 \text{ ft} \quad (487 \text{ ft})$$

The answer is (B).

27. A sketch of the roadways and walkway is shown.

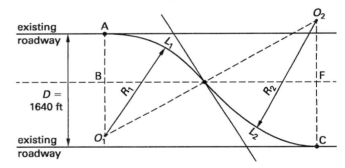

Since the two roadways are parallel, the deflection angles of the two circular curves are equal.

$$I_1 = I_2 = I$$

Also,

$$\text{AB} = R_1(1 - \cos I_1)$$
$$\text{FC} = R_2(1 - \cos I_2)$$
$$D = \text{AB} + \text{FC}$$

Substituting,

$$D = \text{AB} + \text{FC}$$
$$= R_1(1 - \cos I_1) + R_2(1 - \cos I_2)$$
$$= R_1(1 - \cos I) + R_2(1 - \cos I)$$
$$= (R_1 + R_2) - (R_1 + R_2)\cos I$$

Rearrange and solve for deflection angle I.

$$D = (R_1 + R_2) - (R_1 + R_2)\cos I$$

$$I = \arccos \frac{R_1 + R_2 - D}{R_1 + R_2}$$

$$= \arccos \frac{1300 \text{ ft} + 1970 \text{ ft} - 1640 \text{ ft}}{1300 \text{ ft} + 1970 \text{ ft}}$$

$$= 60.1°$$

The length of the first curve is

$$L_1 = \frac{2\pi R_1 I}{360°}$$

$$= \frac{(2\pi)(1300 \text{ ft})(60.1°)}{360°}$$

$$= 1364 \text{ ft}$$

The length of the second curve is

$$L_2 = \frac{2\pi R_2 I}{360°}$$

$$= \frac{(2\pi)(1970 \text{ ft})(60.1°)}{360°}$$

$$= 2066 \text{ ft}$$

The length of the new walkway is

$$L_{\text{new walkway}} = L_1 + L_2$$

$$= 1364 \text{ ft} + 2066 \text{ ft}$$

$$= 3430 \text{ ft}$$

The answer is (C).

28. The station of the BVC is

$$\text{sta}_{\text{BVC}} = V_{\text{sta}} - \frac{L}{2}$$

$$= \text{sta } 142{+}70 - \frac{820 \text{ ft}}{2}$$

$$= \text{sta } 138{+}60$$

The elevation of the BVC is

$$E_{\text{BVC}} = E_V - G_1 \frac{L}{2}$$

$$= 625 \text{ ft} - (0.035)\left(\frac{820 \text{ ft}}{2}\right)$$

$$= 610.7 \text{ ft}$$

The rate of grade change per station is

$$R = \frac{G_2 - G_1}{L}$$

$$= \frac{2.0\% - (-3.5\%)}{8.2 \text{ sta}}$$

$$= 0.67 \text{ \%/sta}$$

The location of the high point is

$$x_{\text{HP}} = \frac{-G_1}{R}$$

$$= \frac{-(-3.5\%)}{0.67 \frac{\%}{\text{sta}}}$$

$$= 522 \text{ ft}$$

The station of the high point is

$$\text{sta}_{\text{HP}} = \text{sta}_{\text{BVC}} + x_{\text{HP}}$$

$$= \text{sta } 138{+}60 + 522 \text{ ft}$$

$$= \text{sta } 143{+}82$$

The elevation of the high point is

$$E_{\text{HP}} = \frac{R x_{\text{HP}}^2}{2L} + G_1 x_{\text{HP}} + E_{\text{BVC}}$$

$$= \frac{\left(-0.055 \frac{\text{ft}}{\text{ft}}\right)(522 \text{ ft})^2}{(2)(820 \text{ ft})}$$

$$\quad + \left(0.035 \frac{\text{ft}}{\text{ft}}\right)(522 \text{ ft}) + 610.7 \text{ ft}$$

$$= 619.8 \text{ ft} \quad (620 \text{ ft})$$

The answer is (C).

29. The change in gradient is

$$A = |G_2 - G_1|$$

$$= |3.7\% - (-2.5\%)|$$

$$= 6.2$$

From Table 3-36 in AASHTO's *A Policy on Geometric Design of Highways and Streets*, the stopping sight distance for a sag vertical curve with a design speed of 35 mph is 250 ft.

If the sight stopping distance, S, is greater than the

length of the curve, L, then the minimum length of the sag vertical curve based on stopping sight distance is

$$
\begin{aligned}
L_{0,\text{sight}} &= 2S - \frac{400 + 3.5S}{A} \\
&= (2)(250 \text{ ft}) - \frac{400 + (3.5)(250 \text{ ft})}{6.2} \\
&= 294.4 \text{ ft}
\end{aligned}
$$

However, $S < L$, so the condition $S > L$ is not satisfied.

If $S < L$, then

$$
\begin{aligned}
L_{1,\text{sight}} &= \frac{AS^2}{400 + 3.5S} \\
&= \frac{(6.2)(250 \text{ ft})^2}{400 + (3.5)(250 \text{ ft})} \\
&= 303.9 \text{ ft}
\end{aligned}
$$

Since $S < L$, the condition is satisfied. The minimum length of the sag vertical curve based on stopping sight distance is 303.9 ft.

The minimum length of a sag vertical curve based on rider comfort is

$$
\begin{aligned}
L_{\text{comfort}} &= \frac{A v_{\text{mph}}^2}{46.5} \\
&= \frac{(6.2)(35 \text{ mph})^2}{46.5} \\
&= 163.3 \text{ ft}
\end{aligned}
$$

The difference in length between the minimum curve length required for stopping sight distance and the minimum curve length required for rider comfort is

$$
\begin{aligned}
\Delta L &= L_{1,\text{sight}} - L_{\text{comfort}} \\
&= 303.9 \text{ ft} - 163.3 \text{ ft} \\
&= 140.6 \text{ ft} \quad (140 \text{ ft})
\end{aligned}
$$

The answer is (C).

30. Find the rate of grade change per station.

$$
\begin{aligned}
R &= \frac{G_2 - G_1}{L} \\
&= \frac{(1.8\% - (-1.3\%))\left(100 \ \frac{\text{ft}}{\text{sta}}\right)}{300 \text{ ft}} \\
&= 1.03 \ \%/\text{sta}
\end{aligned}
$$

The horizontal distance from the BVC to the curve low point is

$$
\begin{aligned}
x &= \frac{-G_1}{R} \\
&= \frac{-(-1.3\%)}{1.03 \ \frac{\%}{\text{sta}}} \\
&= 1.26 \text{ sta} \quad (126 \text{ ft})
\end{aligned}
$$

The elevation of the BVC is

$$
\begin{aligned}
E_{\text{BVC}} &= E_{\text{PVI}} - \frac{LG_1}{2} \\
&= 810 \text{ ft} - \frac{(300 \text{ ft})(-0.013)}{2} \\
&= 811.95 \text{ ft}
\end{aligned}
$$

The elevation of the low point is

$$
\begin{aligned}
E_x &= \frac{Rx^2}{2L} + G_1 x + E_{\text{BVC}} \\
&= \frac{\big(0.018 - (-0.013)\big)(126 \text{ ft})^2}{(2)(300 \text{ ft})} \\
&\quad + (-0.013)(126 \text{ ft}) + 811.95 \text{ ft} \\
&= 811 \text{ ft}
\end{aligned}
$$

The minimum elevation of the overpass is

$$
\begin{aligned}
E_{\text{overpass}} &= E_x + 18 \text{ ft} \\
&= 811 \text{ ft} + 18 \text{ ft} \\
&= 829 \text{ ft}
\end{aligned}
$$

The answer is (C).

31. The total lost time, or the sum of the yellow change and red clearance intervals for all phases, is

$$
\begin{aligned}
L &= \sum\big((\text{yellow change interval}) \\
&\quad + (\text{red clearance interval})\big) \\
&= (4.5 \text{ sec} + 5.0 \text{ sec} + 4.5 \text{ sec}) \\
&\quad + (3.0 \text{ sec} + 4.0 \text{ sec} + 4.5 \text{ sec}) \\
&= 25.5 \text{ sec}
\end{aligned}
$$

The sum of the degree of saturation is

$$
\begin{aligned}
\sum Y_i &= \sum\big((\text{critical volume}) - (\text{capacity ratios})\big) \\
&= \frac{100 \ \frac{\text{veh}}{\text{hr}}}{1900 \ \frac{\text{veh}}{\text{hr}}} + \frac{350 \ \frac{\text{veh}}{\text{hr}}}{1900 \ \frac{\text{veh}}{\text{hr}}} + \frac{250 \ \frac{\text{veh}}{\text{hr}}}{1900 \ \frac{\text{veh}}{\text{hr}}} \\
&= 0.37
\end{aligned}
$$

Using Webster's equation, the optimum cycle length is

$$C = \frac{1.5L + 5}{1 - \sum\limits_{\substack{\text{critical} \\ \text{phases}}} Y_i}$$

$$= \frac{(1.5)(25.5 \text{ sec}) + 5}{1 - 0.37}$$

$$= 68.7 \text{ sec} \quad (70 \text{ sec})$$

The answer is (B).

32. The accident rate for the roadway segment is

$$R = \frac{(\text{no. of accidents})(10^8)}{(\text{ADT})(\text{no. of years})\left(365 \dfrac{\text{day}}{\text{yr}}\right) L_{\text{mi}}}$$

$$= \frac{(16)(10^8)}{\left(12{,}000 \dfrac{\text{veh}}{\text{day}}\right)(2 \text{ yr})\left(365 \dfrac{\text{day}}{\text{yr}}\right)(120 \text{ mi} - 110 \text{ mi})}$$

$$= 18.3 \text{ crashes}/10^8 \text{ veh-mi} \quad (18.3 \text{ crashes per HMVM})$$

The answer is (C).

33. Use the *Manual on Uniform Traffic Control Devices* (MUTCD), Sec. 4C.05 to determine if the location warrants a traffic signal. Warrant 4, Pedestrian Volume, pertains to intersections and midblock locations where heavy street traffic excessively delays pedestrian crossing time. The nearest existing signal is at least 300 ft from the midblock location, so Warrant 4 is applicable. The posted speed limit exceeds 35 mph, and the available data is for four consecutive 15-minute periods, so use criterion B to determine if Warrant 4 is satisfied.

A signal is warranted if the per-hour pedestrian volume is above the curve found in MUTCD Fig. 4C-8.

For the peak morning hour, the total traffic volume is 1323, and the pedestrian volume is 97. The plotted point is above the curve in Fig. 4C-8. The pedestrian volume warrant is satisfied.

For the peak afternoon hour, the total traffic volume is 1291, and the pedestrian volume is 49. The plotted point is below the curve. The pedestrian volume warrant is not satisfied.

The answer is (A).

34. Rearrange the equation for skidding distance to find the velocity of the vehicle when the skidding began.

$$s_b = \frac{\text{v}_1^2 - \text{v}_2^2}{30(f + G)}$$

$$\text{v}_1^2 = \sqrt{30 s_b(f + G) + \text{v}_2^2}$$

$$= \sqrt{(30)(517 \text{ ft})\big(0.30 + (-0.06)\big) + (30)^2}$$

$$= 68 \text{ mph}$$

When its wheels locked up, the vehicle was traveling over the posted speed limit by

$$68 \text{ mph} - 60 \text{ mph} = 8 \text{ mph}$$

The answer is (B).

35. Use the *Highway Capacity Manual* (HCM) to find the heavy-adjustment factor for the freeway segment. The driver population adjustment factor for an urban freeway is $f_p = 1.0$. For an urban multilane freeway, the peak hour factor is PHF = 0.92. For 10% trucks, the truck fractional value is $P_T = 0.10$. Since there are no recreational vehicles (RV), the RV fractional value is $P_R = 0$. From HCM Exh. 11-10, the passenger car equivalent for trucks on a freeway segment on rolling terrain is $E_T = 2.5$. Substituting, the heavy-vehicle adjustment factor is

$$f_{\text{HV}} = \frac{1}{1 + P_T(E_T - 1) + P_R(E_R - 1)}$$

$$= \frac{1}{1 + (0.10)(2.5 - 1) + 0}$$

$$= 0.87$$

From HCM Exh. 11-17, the maximum service flow rate for the freeway segment is $v_p = 1630$ vehicles per hour per lane. Rearrange HCM Eq. 11-2 to find the number of lanes required for LOS C.

$$v_p = \frac{V}{(\text{PHF})\, N f_{\text{HV}} f_p}$$

$$N = \frac{V}{v_p(\text{PHF}) f_{\text{HV}} f_p}$$

$$= \frac{5000 \dfrac{\text{veh}}{\text{hr}}}{\left(1630 \dfrac{\dfrac{\text{veh}}{\text{hr}}}{\text{ln}}\right)(0.92)(0.87)(1.0)}$$

$$= 3.8 \text{ ln} \quad (4 \text{ ln})$$

The answer is (C).

36. From the *Highway Capacity Manual* (HCM), Exh. 11-10;, the truck and RV equivalents for rolling terrain are $E_T = 2.5$ and $E_R = 2.0$, respectively. The proportion of trucks and buses in the traffic stream, P_T, is 0.07. The proportion of RVs in the traffic stream, P_R, is 0.02. Using HCM Eq. 11-3, the heavy-vehicle adjustment factor is

$$f_{\text{HV}} = \frac{1}{1 + P_T(E_T - 1) + P_R(E_R - 1)}$$
$$= \frac{1}{1 + (0.07)(2.5 - 1) + (0.02)(2.0 - 1)}$$
$$= 0.89$$

Using HCM Eq. 11-2, the maximum freeway service flow rate is

$$v_p = \frac{V}{(\text{PHF})\,N f_{\text{HV}} f_p}$$
$$= \frac{3250 \dfrac{\text{veh}}{\text{hr}}}{\ln}$$
$$= \frac{3250 \dfrac{\text{veh}}{\text{hr}}}{(0.87)(3)(0.89)(0.95)}$$
$$= 1473 \text{ pcphpl}$$

The traffic density is

$$D = \frac{v_p}{S}$$
$$= \frac{1473 \dfrac{\dfrac{\text{veh}}{\text{hr}}}{\ln}}{55 \dfrac{\text{mi}}{\text{hr}}}$$
$$= 26.8 \text{ pcpmpl}$$

From HCM Exh. 11-5, a density greater than 26 but less than or equal to 35 indicates LOS D.

The answer is (C).

37. Use the HCM equations for off-street pedestrian facilities to find the pedestrian unit flow rate. The effective width of the walkway, W_E, excludes the decorative planting area.

$$v_p = \frac{v_{15}}{15\,W_E}$$
$$= \frac{25 \text{ ped}}{(15 \text{ min})(10 \text{ ft} - 3 \text{ ft})}$$
$$= 0.238 \text{ ped/min-ft}$$

Using the HCM-recommended pedestrian walking speed

for elderly populations, 3.03 ft/sec, the pedestrian space, A_p, is

$$A_p = \frac{S_{\text{pedestrian}}}{v_p}$$
$$= \frac{\left(3.03 \dfrac{\text{ft}}{\text{sec}}\right)\left(60 \dfrac{\text{sec}}{\text{min}}\right)}{0.238 \dfrac{\text{ped}}{\text{min-ft}}}$$
$$= 764 \text{ ft}^2/\text{ped}$$

HCM Exh. 23-1 states that a per-pedestrian space greater than 60 ft² indicates LOS A.

The answer is (A).

38. Consult MUTCD Sec. 6C-08. From Table 6C-3, the length of a merging taper is at least L. Using the formula in Table 6C-4 for speeds of at least 45 mph, the minimum taper length recommended is

$$L = WS$$
$$= (12 \text{ ft})(55 \text{ mph})$$
$$= 660 \text{ ft}$$

(Units are not consistent in this empirically derived formula.)

The answer is (D).

39. MUTCD Sec. 6E.03, Hand-Signaling Devices, states that STOP/SLOW paddles must be reflectorized when used during the nighttime. Since the paddle is being used at the traffic control zone during the daytime, the paddle does not need to be reflectorized. Option B is correct.

Option A, option C, and option D describe paddle requirements listed in MUTCD Sec. 6E.03.

The answer is (B).

40. Traffic volumes per 15-minute interval and continuous hourly volumes are shown.

interval	traffic volume	continuous hourly volume
1:00–1:15	119	-
1:15–1:30	133	-
1:30–1:45	152	-
1:45–2:00	171	575
2:00–2:15	185	641
2:15–2:30	193	701
2:30–2:45	176	725
2:45–3:00	179	733

The highest continuous hourly volume is 733 vehicles and occurs from 2:00 to 3:00. The peak hour is 2:00–3:00.

Using HCM Eq. 4-2, the peak hour factor is

$$\text{PHF} = \frac{V_{\text{vph}}}{4\,V_{15\text{ min,peak}}}$$

$$= \frac{\left(733\ \dfrac{\text{veh}}{\text{hr}}\right)}{(4)\left(193\ \dfrac{\text{veh}}{\text{hr}}\right)}$$

$$= 0.95$$

The answer is (C).

Solutions
Practice Exam 2

41. Solve for the peak flow by using each answer option to determine which option will yield the estimated flow of 102 ft³/sec.

Calculate the proposed bottom width for a swale 5.6 ft wider at the bottom.

Find the cross-sectional flow area of the channel.

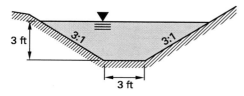

(not to scale)

$$A = \left(b + \frac{h}{\tan\theta}\right)h$$

$$= \left(8.6 \text{ ft} + \frac{3 \text{ ft}}{\dfrac{1}{3}}\right)(3 \text{ ft})$$

$$= 52.8 \text{ ft}^2$$

Calculate the wetted perimeter.

$$P_w = b + 2\left(\frac{h}{\sin\theta}\right)$$

$$= 8.6 \text{ ft} + (2)\left(\frac{3 \text{ ft}}{\dfrac{1}{\sqrt{10}}}\right)$$

$$= 27.6 \text{ ft}$$

Calculate the hydraulic radius.

$$R = \frac{A}{P_w}$$

$$= \frac{52.8 \text{ ft}^2}{27.6 \text{ ft}}$$

$$= 1.9 \text{ ft}$$

The peak flow for a swale 5.6 ft wider at the bottom is

$$Q = \left(\frac{1.49}{n}\right)AR^{2/3}\sqrt{S}$$

$$= \left(\frac{1.49}{0.08}\right)(52.8 \text{ ft}^2)(1.9 \text{ ft})^{2/3}\sqrt{0.003 \frac{\text{ft}}{\text{ft}}}$$

$$= 82.6 \text{ ft}^3/\text{sec}$$

Repeat the same steps for the next option B, a swale 8.8 ft wider at the bottom, which results in the following values

$$A = 62.4 \text{ ft}^2$$

$$P_w = 30.8 \text{ ft}$$

$$R = 2.0 \text{ ft}$$

$$Q = 102 \text{ ft}^3/\text{sec}$$

A swale that is 8.8 ft wider at the bottom has the capacity to carry 102 ft³/sec.

The answer is (B).

42. Per the hydrograph, the pre-development peak flow is 80 ft³/sec.

Identify the post-development volume of runoff to be detained where peak flow cannot exceed 80 ft³/sec. From the post-development hydrograph, the area above 80 ft³/sec represents the runoff volume that needs to be detained.

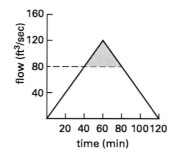

post-development

Calculate the base of the triangle.

$$b = 80 \text{ min} - 40 \text{ min}$$

$$= 40 \text{ min}$$

Calculate the height of the triangle.

$$h = Q_{\text{post-peak}} - Q_{\text{pre-peak}}$$
$$= 120 \ \frac{\text{ft}^3}{\text{sec}} - 80 \ \frac{\text{ft}^3}{\text{sec}}$$
$$= 40 \ \text{ft}^3/\text{sec}$$

Calculate the area of the triangle that represents the volume to be detained.

$$A = \frac{bh}{2}$$
$$= \frac{(40 \ \text{min})\left(40 \ \dfrac{\text{ft}^3}{\text{sec}}\right)\left(60 \ \dfrac{\text{sec}}{\text{min}}\right)}{(2)\left(43{,}560 \ \dfrac{\text{ft}^2}{\text{acre}}\right)}$$
$$= 1.1 \ \text{ac-ft}$$

The answer is (A).

43. Sort the projects in ascending order based on their present value cost estimate.

safety improvement project	present value cost estimate
I. traffic control at Racetrack Blvd.	$1,245,000
II. road widening at Oregon Ave.	$2,346,000
III. roundabout at Tampa Blvd. and Florida Ave.	$3,120,000
IV. traffic signal at Nebraska St. and University Dr.	$4,250,000

Using Eq. 8.3 of the AASHTO *Highway Safety Manual*, compare the first two projects and determine a preferred option.

$$\text{incremental BCR}_{\text{II,I}} = \frac{\text{PV}_{\text{benefits II}} - \text{PV}_{\text{benefits I}}}{\text{PV}_{\text{costs II}} - \text{PV}_{\text{costs I}}}$$
$$= \frac{\$8{,}324{,}000 - \$5{,}438{,}000}{\$2{,}346{,}000 - \$1{,}245{,}000}$$
$$= 2.62$$

Since the incremental BCR is greater than 1, the higher cost project of the road widening at Oregon Ave. is the preferred option of the first two projects, and the base alternative of the traffic control at Racetrack Blvd. is eliminated.

Compare the road widening at Oregon Ave. with the next higher value project, the roundabout at Tampa Blvd. and Florida Ave.

$$\text{incremental BCR}_{\text{III, II}} = \frac{\text{PV}_{\text{benefits III}} - \text{PV}_{\text{benefits II}}}{\text{PV}_{\text{costs III}} - \text{PV}_{\text{costs II}}}$$
$$= \frac{\$7{,}325{,}000 - \$8{,}324{,}000}{\$3{,}120{,}000 - \$2{,}346{,}000}$$
$$= -1.29$$

Since the incremental BCR is less than 1, the base alternative of the road widening at Oregon Ave. is the preferred option, and the higher cost alternative of the roundabout at Tampa Blvd. and Florida Ave. is eliminated.

Compare the road widening at Oregon Ave. with the final project.

$$\text{incremental BCR}_{\text{IV, II}} = \frac{\text{PV}_{\text{benefits IV}} - \text{PV}_{\text{benefits II}}}{\text{PV}_{\text{costs IV}} - \text{PV}_{\text{costs II}}}$$
$$= \frac{\$8{,}385{,}000 - \$8{,}324{,}000}{\$4{,}250{,}000 - \$2{,}346{,}000}$$
$$= 0.03$$

Since the incremental BCR is less than 1, the base alternative of the road widening at Oregon Ave. is the preferred option.

The answer is (D).

44. Use the AASHTO pavement layer thickness equation and rearrange the terms to find the FDR base layer coefficient.

$$\text{SN} = D_1 a_1 + D_2 a_2 m_2 + D_3 a_3 m_3$$
$$a_2 = \frac{\text{SN} - D_1 a_1 - D_3 a_3 m_3}{D_2 m_2}$$
$$= \frac{4.5 - (3 \ \text{in})\left(0.4 \ \dfrac{1}{\text{in}}\right) - (12 \ \text{in})\left(0.1 \ \dfrac{1}{\text{in}}\right)(1.0)}{10 \ \text{in} \ (1.0)}$$
$$= 0.21 \ 1/\text{in}$$

Use the AASHTO resilient modulus equation and rearrange the terms to find the FDR base resilient modulus.

$$a_2 = 0.249 \ \log_{10} E_{\text{BS}} - 0.977$$
$$E_B = 10^{(a_2 + 0.977)/0.249}$$
$$= \frac{10^{((0.21 \ 1/\text{in}) + 0.977)/0.249}}{1000 \ \dfrac{\text{psi}}{\text{ksi}}}$$
$$= 58.5 \ \text{ksi}$$

From the provided graph, the percentage of fly ash needed for a base resilient modulus of 58.5 ksi is 5%.

The answer is (D).

45. Calculate the curve internal angle, I.

$$I = 180° - (15° \, 19' \, 15'' + 12° \, 40' \, 45'')$$
$$= 152°$$

Rearrange the terms in the arc length equation, and calculate the curve radius.

$$L = \frac{2\pi R I}{360°}$$
$$R = \frac{L(360°)}{2\pi I}$$
$$= \frac{(376 \text{ ft})(360°)}{(2\pi)(152°)}$$
$$= 141.7 \text{ ft}$$

Find the tangent length, T.

$$T = R \tan \frac{I}{2}$$
$$= (141.7 \text{ ft})\left(\tan \frac{152°}{2}\right)$$
$$= 568.3 \text{ ft}$$

The answer is (A).

46. Calculate the queue service time.

$$g_s = \frac{Q_r}{\text{queue discharge rate}}$$
$$= \frac{18 \text{ veh}}{0.8 \frac{\text{veh}}{\text{sec}}}$$
$$= 22.5 \text{ sec}$$

Determine the green extension time from HCM Eq. 18.13 and Exh. 18.14.

$$D_p = l_1 + g_s + g_e + Y + R_C$$
$$g_e = D_p - l_1 - g_s - Y - R_C$$
$$= 45 \text{ sec} - 2 \text{ sec} - 22.5 \text{ sec} - 2.5 \text{ sec} - 1 \text{ sec}$$
$$= 17.0 \text{ sec}$$

The answer is (B).

47. The stopping sight distance (SSD) is the distance traveled during the brake reaction time plus the braking distance.

$$\text{SSD} = d_{\text{reaction}} + d_{\text{braking}}$$

Calculate the brake reaction distance.

$$d_{\text{reaction}} = 1.47 v_{\text{mph}} t = (1.47)(45 \text{ mph})(1.5 \text{ sec})$$
$$= 99.2 \text{ ft}$$

Using Eq. 3-3 in the AASHTO *Green Book*, calculate the braking distance on a 4% uphill grade (scenario I).

$$d_{\text{braking}} = \frac{v_{\text{mph}}^2}{30(f + G)}$$
$$= \frac{(45 \text{ mph})^2}{(30)\left(0.32 + 0.04 \frac{\text{ft}}{\text{ft}}\right)}$$
$$= 187.5 \text{ ft}$$

Use the reaction and braking values to calculate the stopping sight distance.

$$\text{SSD} = 99.2 \text{ ft} + 187.5 \text{ ft} = 286.7 \text{ ft}$$

Because 325 ft (the distance of the object in the roadway) is greater than 286.7 ft, in scenario I the driver has sufficient distance to stop the vehicle.

Now use Eq. 3-3 in the AASHTO *Green Book* to calculate the braking distance on a −4% downhill grade.

$$d_{\text{braking}} = \frac{v_{\text{mph}}^2}{30(f + G)}$$
$$= \frac{(45 \text{ mph})^2}{(30)\left(0.32 - 0.04 \frac{\text{ft}}{\text{ft}}\right)}$$
$$= 241.07 \text{ ft}$$

Use the reaction and braking values to calculate the stopping sight distance.

$$\text{SSD} = 99.2 \text{ ft} + 241.07 \text{ ft} = 340.27 \text{ ft}$$

Because 325 ft (the distance of the object in the roadway) is less than 340.27 ft, in scenario II the driver does not have sufficient distance to stop the vehicle.

The answer is (A).

48. Refer to MUTCD Table 2C-4.

The Signal Ahead warning sign is considered as condition B, with a deceleration to 0 mph. With a posted speed limit of 55 mph, the sign should be placed 325 ft in advance of the traffic signal.

The Turn warning sign is considered as condition B. With an advisory speed of 20 mph, the sign should be placed 100 ft in advance of the turn.

The answer is (A).

49. Determine the station at PC_1.

$$\begin{aligned} \text{sta } PC_1 &= \text{sta A} + \text{segment}_1 \\ &= \text{sta } 14{+}25 + 225 \text{ ft} \\ &= \text{sta } 16{+}50 \end{aligned}$$

Calculate the length of curve 1.

$$\begin{aligned} L_1 &= \frac{2\pi R_1 \Delta_1}{360°} \\ &= \frac{(2\pi)(400 \text{ ft})(50° \, 58' \, 32'')}{360°} \\ &= 355.9 \text{ ft} \end{aligned}$$

Determine the station at PT_1.

$$\begin{aligned} \text{sta } PT_1 &= \text{sta } PC_1 + L_1 \\ &= \text{sta } 16{+}50 + 355.9 \text{ ft} \\ &= \text{sta } 20{+}06 \end{aligned}$$

Determine the station at PC_2.

$$\begin{aligned} \text{sta } PC_2 &= \text{sta } PT_1 + \text{segment}_2 \\ &= \text{sta } 20{+}06 + 300 \text{ ft} \\ &= \text{sta } 23{+}06 \end{aligned}$$

Calculate the length of curve 2.

$$\begin{aligned} L_2 &= \frac{2\pi R_2 \Delta_2}{360°} \\ &= \frac{(2\pi)(425 \text{ ft})(45°)}{360°} \\ &= 333.8 \text{ ft} \end{aligned}$$

Determine the station at PT_2.

$$\begin{aligned} \text{sta } PT_2 &= \text{sta } PC_2 + L_2 \\ &= \text{sta } 23{+}06 + 333.8 \text{ ft} \\ &= \text{sta } 26{+}40 \end{aligned}$$

The answer is (D).

50. For situation I, use the AASHTO *Green Book* Table 9-7, to find the time gap for a combination truck turning right from a stop.

$$t_g = 10.5 \text{ sec}$$

Calculate the intersection sight distance.

$$\begin{aligned} \text{ISD} &= 1.47 \, V_{\text{major}} t_g \\ &= (1.47)(45 \text{ mph})(10.5 \text{ sec}) \\ &= 695 \text{ ft} \end{aligned}$$

For situation II, use the AASHTO *Green Book* Table 9-3, to determine the intersection sight distance for intersections with no traffic controls and vehicles traveling at 80 mph.

$$\text{ISD} = 485 \text{ ft}$$

Use the AASHTO *Green Book* Table 9-4, to adjust the intersection sight distance due to the -6% approach grade at 80 mph.

$$\begin{aligned} \text{ISD} &= (485 \text{ ft})(1.2) \\ &= 582 \text{ ft} \end{aligned}$$

For situation III, use the AASHTO *Green Book* Table 9-5, to determine the time gap for a passenger car turning left from a stop.

$$t_g = 7.5 \text{ sec}$$

Calculate the intersection sight distance.

$$\begin{aligned} \text{ISD} &= 1.47 \, V_{\text{major}} t_g \\ &= (1.47)(55 \text{ mph})(7.5 \text{ sec}) \\ &= 606 \text{ ft} \end{aligned}$$

For situation IV, use the AASHTO *Green Book* Table 9-7, to determine the time gap for a single-unit truck making a crossing maneuver from a stop.

$$\begin{aligned} t_g &= 8.5 \text{ sec} + 0.7 \text{ sec} + 0.7 \text{ sec} \\ &= 9.9 \text{ sec} \end{aligned}$$

Calculate the intersection sight distance.

$$\begin{aligned} \text{ISD} &= 1.47 \, V_{\text{major}} t_g \\ &= (1.47)(45 \text{ mph})(9.9 \text{ sec}) \\ &= 655 \text{ ft} \end{aligned}$$

Situation I requires the longest intersection sight distance (695 ft).

The answer is (A).

51. The assumption is that the headlight sight distance, S, should be less than the length of the vertical curve, L. Calculate the length of the vertical curve.

$$\begin{aligned} L &= \frac{AS^2}{400 + 3.5S} \\ &= \frac{\big(3\% - (-2\%)\big)(400 \text{ ft})^2}{400 + (3.5)(400 \text{ ft})} \\ &= 444 \text{ ft} \quad (450 \text{ ft}) \end{aligned}$$

Since $S < L$, the condition is met and the equation is applicable.

The answer is (C).

52. Determine the elevation at the beginning of the vertical curve.

$$\text{elev}_{BVC} = \text{elev}_Q - (\text{sta BVC} - \text{sta Q})G_1$$
$$= 210.6 \text{ ft} + (\text{sta } 15{+}75 - \text{sta } 13{+}55)$$
$$\times \left(-0.02 \frac{\text{ft}}{\text{ft}}\right)$$
$$= 206.2 \text{ ft}$$

Calculate the rate of grade change per station.

$$R = \frac{G_2 - G_1}{L}$$
$$= \frac{-0.0475 \frac{\text{ft}}{\text{ft}} - \left(-0.02 \frac{\text{ft}}{\text{ft}}\right)}{\text{sta } 19{+}75 - \text{sta } 15{+}75}$$
$$= -6.88 \times 10^{-5} \text{ 1/ft}$$

Calculate the distance from the BVC to point B.

$$x = \text{sta B} - \text{sta BVC}$$
$$= \text{sta } 17{+}00 - \text{sta } 15{+}75$$
$$= 125 \text{ ft}$$

Determine the elevation of point B.

$$\text{elev}_B = \frac{Rx^2}{2} + G_1 x + \text{elev}_{BVC}$$
$$= \frac{(6.88 \times 10^{-5})(125 \text{ ft})^2}{2}$$
$$+ \left(-0.02 \frac{\text{ft}}{\text{ft}}\right)(125 \text{ ft}) + 206.2 \text{ ft}$$
$$= 204.2 \text{ ft}$$

The answer is (B).

53. From the AASHTO *Green Book* Table 9-22, the desirable full deceleration length for a deceleration lane with a speed of 50 mph is 425 ft.

Calculate the desirable taper distance.

$$L_{\text{decel}} = L_2 + L_3$$

Since $L_2 = L_3$,

$$L_2 = \frac{L_{\text{decel}}}{2}$$
$$= \frac{425 \text{ ft}}{2}$$
$$= 213 \text{ ft} \quad (215 \text{ ft})$$

The answer is (A).

54. Determine the average hourly volume for the major street.

$$V_{\text{vph,major}} = \frac{V_{\text{major}}}{8 \text{ hr}}$$
$$= \frac{6000 \text{ veh}}{8 \text{ hr}}$$
$$= 750 \text{ vph}$$

Determine the average hourly volume for the minor street.

$$V_{\text{vph,minor}} = \frac{(0.60) V_{\text{minor}}}{8 \text{ hr}}$$
$$= \frac{(0.60)(1600 \text{ veh})}{8 \text{ hr}}$$
$$= 120 \text{ vph}$$

From MUTCD Table 4C-1, condition A, 80% column, 2 lane major/1 lane minor road, the average vehicle volume needs to be at least 480 vph on the major street and 120 vph on the minor street to satisfy the condition. The volume on the major street and on the minor street meets the minimum requirement, so condition A is met for the 80% column.

From MUTCD Table 4C-1, condition B, 80% column, 2 lane major/1 lane minor road, the average vehicle volume needs to be at least 720 vph on the major street and 60 vph on the minor street to satisfy the condition. The volume on the major street and on the minor street meets the minimum requirement, so condition B is met for the 80% column.

Conditions A and B have both been met for the 80% column.

The answer is (C).

55. Convert the purchase price to an annual cost for machine 1.

$$A_{1A/P} = P \frac{i(1 + i)^n}{(1 + i)^n - 1}$$
$$= (\$18{,}000)\left(\frac{(0.08)(1 + 0.08)^5}{(1 + 0.08)^5 - 1}\right)$$
$$= \$4508$$

Convert the salvage value to an annual cost for machine 1.

$$A_{1A/F} = F\frac{i}{(1+i)^n - 1}$$
$$= (\$7500)\left(\frac{0.08}{(1+0.08)^5 - 1}\right)$$
$$= \$1279$$

Calculate the annual costs for machine 1.

$$A_1 = A_{1A/P} + A_{1M} - A_{1A/F}$$
$$= \$4508 + \$500 - \$1279$$
$$= \$3729$$

Convert the purchase price to an annual cost for machine 2.

$$A_{2A/P} = P\frac{i(1+i)^n}{(1+i)^n - 1}$$
$$= (\$15,000)\left(\frac{(0.08)(1+0.08)^5}{(1+0.08)^5 - 1}\right)$$
$$= \$3757$$

Convert the salvage value to an annual cost for machine 2.

$$A_{2A/F} = F\frac{i}{(1+i)^n - 1}$$
$$= (\$6000)\left(\frac{0.08}{(1+0.08)^5 - 1}\right)$$
$$= \$1023$$

Calculate the annual costs for machine 2.

$$A_2 = A_{2A/P} + A_{2M} - A_{2A/F}$$
$$= \$3757 + \$700 - \$1023$$
$$= \$3434$$

Calculate the annual savings of choosing machine 1 instead of machine 2.

$$\text{Annual savings} = A_1 - A_2$$
$$= \$3729 - \$3434$$
$$= \$295$$

The answer is (D).

56. A traffic impact study typically includes data for traffic counts, peak hours, travel forecasting, and trip generations.

The answer is (B).

57. According to the AASHTO Exh. 3-1, the probability of a pedestrian fatality when a pedestrian is struck by a vehicle travelling at 20 mph is about 15%, and the probability of a pedestrian fatality when a pedestrian is struck by a vehicle travelling at 40 mph is about 85%.

Calculate the increase in the probability of a pedestrian fatality.

$$x = \frac{p_{40\text{ mph}}}{p_{20\text{ mph}}}$$
$$= \frac{0.85}{0.15}$$
$$= 5.7$$

The answer is (D).

58. Find the length of the curve.

$$\text{sta PT} = \text{sta PC} + L$$
$$L = \text{sta PT} - \text{sta PC}$$
$$= (\text{sta }164 + 00 - \text{sta }150 + 00)\left(100\ \frac{\text{ft}}{\text{sta}}\right)$$
$$= 1400\text{ ft}$$

From the equation for curve length, calculate the radius of the curve.

$$L = \frac{2\pi R I}{360°}$$
$$R = \frac{(1400\text{ ft})(360°)}{(80°)(2\pi)}$$
$$= 1003\text{ ft}$$

For a speed, v, between 50 mph and 70 mph, determine the side friction factor, f_s.

$$f_s = 0.14 - \frac{(0.02)(\text{v}_{\text{mph}} - 50)}{10}$$
$$= 0.14 - \frac{(0.02)(50\text{ mph} - 50\text{ mph})}{10}$$
$$= 0.14$$

Calculate the superelevation.

$$e = \frac{\text{v}^2}{15R} - f_s$$
$$= \frac{(50\text{ mph})^2}{(15)(1003\text{ ft})} - 0.14$$
$$= 0.026 \quad (2.6\%)$$

The answer is (B).

59. Calculate the change in gradient.

$$A = |G_2 - G_1|$$
$$= |-1\% - 3\%|$$
$$= 4\%$$

According to AASHTO, the height of the driver's eye above the road is 3.5 ft.

Calculate the length of the vertical curve assuming the stopping sight distance is less than the length of the vertical curve, $S < L$.

$$L = \frac{AS^2}{(200)\left(\sqrt{h_1} + \sqrt{h_2}\right)^2}$$
$$= \frac{(4\%)(520 \text{ ft})^2}{(200)\left(\sqrt{3.5 \text{ ft}} + \sqrt{2.0 \text{ ft}}\right)^2}$$
$$= 501 \text{ ft}$$

However, $S = 520$ ft is not less than $L = 501$ ft, so the assumption is invalid.

Calculate the length of the vertical curve assuming the length of the vertical curve is less than the stopping sight distance, $S > L$.

$$L = 2S - \frac{(200)\left(\sqrt{h_1} + \sqrt{h_2}\right)^2}{A}$$
$$= (2)(520 \text{ ft}) - \frac{(200)\left(\sqrt{3.5 \text{ ft}} + \sqrt{2.0 \text{ ft}}\right)^2}{(4\%)}$$
$$= 500 \text{ ft}$$

$S = 520$ ft is greater than $L = 500$ ft, so the assumption is valid and $L = 500$ ft.

Calculate the rate of change in grade per station.

$$R = \frac{G_2 - G_1}{L} = \frac{-0.01 - 0.03}{500 \text{ ft}}$$
$$= -0.00008 \text{ 1/ft}$$

Determine the elevation at a distance of ⅓ the curve length from the BVC.

$$\text{elev}_x = \left(\frac{R}{2}\right)x^2 + G_1 x + \text{elev}_{\text{BVC}}$$
$$= \left(\frac{-0.00008 \frac{1}{\text{ft}}}{2}\right)\left(\frac{500 \text{ ft}}{3}\right)^2 + \left(0.03 \frac{\text{ft}}{\text{ft}}\right)\left(\frac{500 \text{ ft}}{3}\right) + 920 \text{ ft}$$
$$= 924 \text{ ft}$$

The answer is (A).

60. Calculate the pedestrian unit flow rate. The HCM recommends a general walking speed of 4.0 ft/sec, and for LOS C, the pedestrian space must be greater than 24 ft^2/ped.

$$v_{\text{ped}} = \frac{S_{\text{ped}}}{A_{\text{ped}}}$$
$$= \frac{4.0 \frac{\text{ft}}{\text{sec}}}{24 \frac{\text{ft}^2}{\text{ped}}}$$
$$= 0.167 \text{ ped/ft-sec}$$

Calculate the minimum effective width of the walkway.

$$v_{\text{ped}} = \frac{v_{15}}{15 W_E}$$
$$W_E = \frac{v_{15}}{15 v_{\text{ped}}}$$
$$= \frac{1200 \text{ ped}}{(15 \text{ min})\left(60 \frac{\text{sec}}{\text{min}}\right)\left(0.167 \frac{\text{ped}}{\text{ft-sec}}\right)}$$
$$= 8 \text{ ft}$$

The answer is (A).

61. The PHF is determined by finding the hour with the largest total volume and then the peak 15 min volume within that hour.

Data inspection indicates that the peak hour is from 5:00 p.m. to 6:00 p.m. The traffic volume for this period is summarized in the table shown. The total hourly volume is 2501 vehicles.

time interval	left	through	right	total
5:00–5:15 p.m.	31	525	61	617
5:15–5:30 p.m.	35	536	75	646
5:30–5:45 p.m.	31	542	70	643
5:45–6:00 p.m.	25	515	55	595
total	122	2118	261	2501

peak hour volumes

Calculate the PHF from the HCM.

$$\text{PHF} = \frac{V}{4 V_{15}}$$

V_{15} is equal to the largest 15 min volume for the

approach within the peak hour, which is 646 vehicles from 5:15 to 5:30 p.m.

$$\text{PHF} = \frac{2501 \text{ vehicles}}{(4)(646 \text{ vehicles})}$$
$$= 0.968 \quad (0.97)$$

The answer is (C).

62. Utilize the *Highway Capacity Manual* (HCM) Eq. 18-16 to determine the volume-to-capacity ratio for the left turn movement.

$$\frac{v}{c} = \frac{v}{s\left(\dfrac{g}{C}\right)}$$

$$= \frac{274 \dfrac{\text{veh}}{\text{hr}}}{\left(1800 \dfrac{\text{veh}}{\text{hr}}\right)\left(\dfrac{12.5 \text{ sec}}{70 \text{ sec}}\right)}$$

$$= 0.852$$

Calculate the uniform control delay, d_1.

$$d_1 = \frac{0.5 C\left(1 - \dfrac{g}{C}\right)^2}{1 - \left(\dfrac{v}{c}\right)\left(\dfrac{g}{C}\right)}$$

$$= \frac{(0.5)(70 \text{ sec})\left(1 - \dfrac{12.5 \text{ sec}}{70 \text{ sec}}\right)^2}{1 - (0.852)\left(\dfrac{12.5 \text{ sec}}{70 \text{ sec}}\right)}$$

$$= 27.9 \text{ sec}$$

The control delay is the sum of the uniform delay, the incremental delay, and the initial queue delay. Calculate the control delay.

$$d = d_1(\text{PF}) + d_2 + d_3$$
$$= (27.9 \text{ sec})(1) + 27.0 \text{ sec} + 0 \text{ sec}$$
$$= 54.9 \text{ sec}$$

For a control delay of 54.9 sec, HCM Exh. 18-4 indicates LOS D for this lane group.

The answer is (D).

63. MUTCD Sec. 6C.03, "Components of Temporary Traffic Control Zones," states that most traffic control zones are divided into four areas (shown in MUTCD Fig. 6C-1): advance warning, transition, activity, and termination.

The answer is (C).

64. Per RDG Fig. 5-2(b), the only situation listed that falls into the "more preferable" area for embankment barrier consideration is a 1.5:1 slope with a 15 ft embankment height and an ADT of 1200 vpd.

The answer is (C).

65. Table 3-1 in the AASHTO *Roadside Design Guide* (RDG) indicates that the clear zone distance, L_C, should be 16–18 ft.

AASHTO Table 3-2 provides a horizontal curve adjustment factor of 1.3, K_{CZ}, for a curve with a 1200 ft radius and a 50 mph design speed.

Compute the minimum clear zone distance.

$$\text{C}_{ZC} = L_C K_{CZ}$$
$$= (16 \text{ ft})(1.3)$$
$$= 20.8 \text{ ft} \quad (21 \text{ ft})$$

Note that the RDG recommends a clear zone distance of 16–18 ft, but the question is asking for the minimum value in the range.

The answer is (C).

66. Per AASHTO GDHS Sec. 4.3, a vehicle lane for a high-speed, high-volume roadway is predominantly 12 ft wide. Per GDHS Sec. 4.4.2, the preferred shoulder width is 10 ft.

$$W_{ROW} = (4 \text{ lanes})\left(12 \dfrac{\text{ft}}{\text{lane}}\right) + (2 \text{ shoulders})\left(10 \dfrac{\text{ft}}{\text{shoulder}}\right)$$
$$= 68 \text{ ft}$$

The answer is (C).

67. The rate of grade change per station is

$$R = \frac{G_2 - G_1}{L}$$

$$= \frac{4.2\% - (-5.0\%)}{\left(\dfrac{160 \text{ ft}}{100 \dfrac{\text{ft}}{\text{sta}}}\right)}$$

$$= 5.75 \ \%/\text{sta}$$

The horizontal distance from the BCV to the low point of the curve is

$$x_{\text{LP}} = \frac{-G_1}{R}$$

$$= \frac{-(-5.0\%)}{\left(\dfrac{5.75 \dfrac{\%}{\text{sta}}}{100 \dfrac{\text{ft}}{\text{sta}}}\right)}$$

$$= 87 \text{ ft}$$

The elevation at the low point of the curve is

$$\text{elev}_{\text{LP}} = \frac{Rx_{\text{LP}}^2}{2} + G_1 x_{\text{LP}} + \text{elev}_{\text{BVC}}$$

$$= \frac{\left(\dfrac{0.0575}{\text{sta}}\right)(87 \text{ ft})^2}{\left(100 \dfrac{\text{ft}}{\text{sta}}\right)(2)} + (-0.05)(87 \text{ ft}) + 513.15 \text{ ft}$$

$$= 511 \text{ ft}$$

The minimum elevation of the bridge soffit is

$$\text{elev}_{\text{bridge}} = \text{elev}_{\text{LP}} + h_{\min}$$

$$= 511 \text{ ft} + 20 \text{ ft}$$

$$= 531 \text{ ft}$$

The answer is (C).

68. The saturation flow rate is

$$s = \frac{3600 \dfrac{\text{sec}}{\text{hr}}}{\text{saturation headway}_{\text{sec/veh}}}$$

$$= \frac{3600 \dfrac{\text{sec}}{\text{hr}}}{2.2 \dfrac{\text{sec}}{\text{veh}}}$$

$$= 1636 \text{ vph}$$

The through-movement capacity is

$$c = Ns\left(\frac{g}{C}\right)$$

$$= (1 \text{ lane})\left(\frac{1636 \dfrac{\text{veh}}{\text{hr}}}{1 \text{ lane}}\right)\left(\frac{29 \text{ sec}}{60 \text{ sec}}\right)$$

$$= 791 \text{ vph}$$

The answer is (C).

69. Calculate L, using MUTCD Table 6C-4, for $S = 45$ mph.

$$L = WS$$

$$= \left(16 \dfrac{\text{ft}}{\text{mph}}\right)(45 \text{ mph})$$

$$= 720 \text{ ft}$$

From MUTCD Table 6C-3, the distance for shifting tapers is $0.5L$.

$$L_{\text{shifting}} = 0.5L$$

$$= (0.5)(720 \text{ ft})$$

$$= 360 \text{ ft}$$

The answer is (D).

70. Providing for all movements to and from the freeway discourages intentional wrong-way entry. Narrowing the arterial highway median opening reduces the probability of left-turn movements onto freeway off-ramps.

Using conventional, easily recognizable interchange patterns reduce wrong-way entries by reducing driver confusion.

Adjacent on- and off-ramps that join a minor road should be separated, not merged.

The answer is (A).

71. The total travel time between the two nodes is

$$t = t_0\left[1 + a\left(\frac{v}{c}\right)^b\right]$$

$$= (47 \text{ min})\left[1 + (0.15)\left(\frac{2200 \dfrac{\text{veh}}{\text{hr}}}{2400 \dfrac{\text{veh}}{\text{hr}}}\right)^{4.0}\right]$$

$$= 52 \text{ min}$$

The answer is (C).

72. The value of the automobile mode is

$$v_{\text{automobile}} = -0.006 t_{\text{automobile}} - 0.004 c_{\text{automobile}}$$
$$- 0.003 W_{\text{automobile}}$$

$$= (-0.006)(45 \text{ min}) - (0.004)(\$8.00)$$
$$- (0.003)(0 \text{ min})$$

$$= -0.302$$

The value of the carpool mode is

$$
\begin{aligned}
v_{\text{carpool}} &= -0.006t_{\text{carpool}} - 0.004c_{\text{carpool}} \\
&\quad - 0.003\,W_{\text{carpool}} \\
&= (-0.006)(55 \text{ min}) - (0.004)(\$4.00) \\
&\quad - (0.003)(10 \text{ min}) \\
&= -0.376
\end{aligned}
$$

The value of the bus mode is

$$
\begin{aligned}
v_{\text{bus}} &= -0.006t_{\text{bus}} - 0.004c_{\text{bus}} \\
&\quad - 0.003\,W_{\text{bus}} \\
&= (-0.006)(65 \text{ min}) - (0.004)(\$2.00) \\
&\quad - (0.003)(20 \text{ min}) \\
&= -0.458
\end{aligned}
$$

The probability that any commuter sampled will commute by bus is

$$
\begin{aligned}
p_{\text{bus}} &= \frac{e^{v_m}}{\sum e^{v_m}} = \frac{e^{-0.458}}{e^{-0.302} + e^{-0.376} + e^{-0.458}} \\
&= 0.307
\end{aligned}
$$

The number of commuters expected to commute by bus is

$$
\begin{aligned}
T &= p_{\text{bus}}N = (0.307)(10{,}000) \\
&= 3070
\end{aligned}
$$

The answer is (A).

73. The sample size is $n = 8$. The sample mean is

$$
\begin{aligned}
\overline{x} &= \frac{\sum x_i}{n} = \frac{(6 + 2 + 7 + 8 + 9 + 11 + 6 + 5)}{8} \\
&= 6.75
\end{aligned}
$$

The standard deviation is

$$
\begin{aligned}
s &= \sqrt{\frac{\sum x_i^2 - n\overline{x}^2}{n-1}} = \sqrt{\frac{416 - (8)(6.75)^2}{8-1}} \\
&= 2.71
\end{aligned}
$$

The standard error for the data set is

$$
\begin{aligned}
s_{\overline{x}} &= \frac{s}{\sqrt{n}} = \frac{2.71}{\sqrt{8}} \\
&= 0.958 \quad (0.96)
\end{aligned}
$$

The answer is (C).

74. Since 25% of crashes are rear-end type, the number of rear-end crashes before implementing the countermeasure is

$$
\begin{aligned}
N_{\text{base}} &= (0.25)\left(264 \; \frac{\text{crashes}}{\text{yr}}\right) \\
&= 66 \text{ crashes/yr}
\end{aligned}
$$

The number of rear-end crashes after implementing the countermeasure is

$$
\begin{aligned}
N_{\text{modified}} &= N_{\text{base}} - N_{\text{reduction}} = 66 \; \frac{\text{crashes}}{\text{yr}} - 45 \; \frac{\text{crashes}}{\text{yr}} \\
&= 21 \text{ crashes/yr}
\end{aligned}
$$

Using the AASHTO *Highway Safety Manual*, the crash modification factor is

$$
\begin{aligned}
\text{CMF} &= \frac{N_{\text{modified}}}{N_{\text{base}}} = \frac{21 \; \dfrac{\text{crashes}}{\text{yr}}}{66 \; \dfrac{\text{crashes}}{\text{yr}}} \\
&= 0.32
\end{aligned}
$$

The crash reduction factor for the countermeasure is

$$
\begin{aligned}
\text{CRF} &= 1 - \text{CMF} = 1 - 0.32 \\
&= 0.68
\end{aligned}
$$

The answer is (B).

75. The control delay for the intersection is

$$
\begin{aligned}
d &= \frac{\sum d_i v_i}{v_i} \\
&= \frac{\begin{array}{c} (12.0 \text{ sec})\left(472 \; \dfrac{\text{veh}}{\text{hr}}\right) + (13.5 \text{ sec})\left(151 \; \dfrac{\text{veh}}{\text{hr}}\right) \\ + (14.0)\left(490 \; \dfrac{\text{veh}}{\text{hr}}\right) \end{array}}{472 \; \dfrac{\text{veh}}{\text{hr}} + 151 \; \dfrac{\text{veh}}{\text{hr}} + 490 \; \dfrac{\text{veh}}{\text{hr}}} \\
&= 13.1 \text{ sec}
\end{aligned}
$$

From Exh. 20-2 in the *Highway Capacity Manual*, a control delay of 13.1 sec indicates LOS B.

The answer is (B).

76. According to the AASHTO *Guide for Planning and Design of Pedestrian Facilities*, a level landing is required for every rise in elevation of 30 in. To be ADA compliant, the number of level landings the pedestrian ramp requires is

$$\text{landings} = \frac{\Delta \text{elevation}}{30 \text{ in}} = \frac{17.5 \text{ ft}}{\dfrac{30 \text{ in}}{12 \dfrac{\text{in}}{\text{ft}}}} = 7$$

The answer is (D).

77. Use the AASHTO GDPS-4-M.

The ESALs for truck A is

$$
\begin{aligned}
\text{ESAL}_A &= \sum (N_{\text{axles}}) \text{LEF} \\
&= (1 \text{ axle})\left(0.623 \frac{\text{ESAL}}{\text{axle}}\right) + (1 \text{ axle})\left(0.005 \frac{\text{ESAL}}{\text{axle}}\right) \\
&= 0.628 \text{ ESAL}
\end{aligned}
$$

The ESALs for truck B is

$$
\begin{aligned}
\text{ESAL}_B &= \sum (N_{\text{axles}}) \text{LEF} \\
&= (2 \text{ axles})\left(0.857 \frac{\text{ESAL}}{\text{axle}}\right) + (1 \text{ axle})\left(0.360 \frac{\text{ESAL}}{\text{axle}}\right) \\
&= 2.07 \text{ ESAL}
\end{aligned}
$$

Compare the loading between truck A and truck B to find the number of passes of truck A equivalent to a single pass of truck B.

$$
\begin{aligned}
x &= \frac{\text{ESAL}_B}{\text{ESAL}_A} = \frac{2.07 \text{ ESAL}}{0.628 \text{ ESAL}} \\
&= 3.30
\end{aligned}
$$

The answer is (B).

78. Corresponding to the peak of the curve, the optimum water content is 13%.

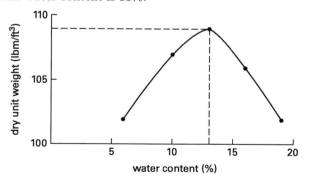

The answer is (C).

79. Signal warrant 3, peak hour, is primarily based on a delay to minor movements during peak hour conditions. MUTCD Sec. 4C.04 gives guidance. The need for a traffic control signal should be considered if the criteria in either of the following two categories are met.

Category A

If all three of the following conditions exist for the same one hour (any four consecutive 15 min periods) of an average day

(1) the total stopped time delay experienced by the traffic on one minor-street approach (one direction only) controlled by a stop sign equals or exceeds four vehicle-hours for a one-lane approach; and

(2) the volume on the same minor-street approach (one direction only) equals or exceeds 100 vehicles per hour for one moving lane of traffic; and

(3) the total entering volume serviced during the hour equals or exceeds 800 vehicles per hour for intersections with four or more approaches

For category A, the first condition is not met for either the a.m. or p.m. peak hour periods. Since all three conditions are not met, category A does not satisfy the warrant for the a.m. or p.m. peak hour period. Category A does not fulfill the conditions for a signal warrant.

Next, investigate category B for a.m. and p.m. peak hours.

Category B

The plotted point representing the vehicles per hour on the major street (total of both approaches) and the corresponding vehicles per hour on the higher-volume minor-street approach (one direction only) for one hour (any four consecutive 15 min periods) of an average day falls above the applicable curve in MUTCD Fig. 4C-3 for the existing combination of approach lanes.

For the a.m. peak hour, the total volume of both approaches on the major-street (NB + SB) is equal to the sum of the 15 min volumes.

$$45 + 60 + 62 + 70 + 105 + 108 + 90 + 110 = 650 \text{ veh/hr}$$

The sum of the 15 min volumes for the higher-volume minor-street (one direction only, EB) approach is

$$15 + 30 + 10 + 20 = 75 \text{ veh/hr}$$

The plotted point for these two values falls well below the applicable curve in MUTCD Fig. 4C-3 (1-lane & 1-lane). Therefore, the a.m. peak hour volume does not satisfy the warrant.

For the p.m. peak hour, the total volume of both approaches on the major-street (NB + SB) is equal to the sum of the 15 min volumes.

$$82 + 96 + 120 + 138 + 71 + 82 + 80 + 85 = 754 \text{ veh/hr}$$

The sum of the 15 min volumes for the higher-volume minor-street (one direction only, EB) approach is

$$59 + 62 + 67 + 64 = 252 \text{ veh/hr}$$

The plotted point for these two values falls just below the applicable curve in MUTCD Fig. 4C-3 (1-lane & 1-lane). Therefore, the p.m. peak hour volume does not satisfy the warrant.

The answer is (D).

80. Calculate the maximum specific gravity for mix design A.

$$G_{mm} = \frac{100\%}{\dfrac{P_s}{G_{se}} + \dfrac{P_b}{G_b}} = \frac{100\%}{\dfrac{95\%}{2.5} + \dfrac{5\%}{2.3}}$$
$$= 2.49$$

Determine the spread rate of mix design A.

$$\text{spread rate} = tG_{mm}\left(43.3 \ \frac{\text{lbm}}{\text{in-yd}^2}\right)$$
$$= (2 \text{ in})(2.49)\left(43.3 \ \frac{\text{lbm}}{\text{in-yd}^2}\right)$$
$$= 216 \text{ lbm/yd}^2$$

The spread rate for mix design A is 216 lbm/yd^2, which is lower than the spread rate estimate of 220 lbm/yd^2.

Use the same calculations to determine the specific gravity and spread rate for the remaining mix designs.

mix design	percent asphalt	percent aggregate	specific gravity of asphalt	specific gravity of aggregate	specific gravity of mix	spread rate for mix (lbm/yd^2)
A	5	95	2.3	2.5	2.49	216
B	6	94	2.4	2.5	2.49	216
C	7	93	2.2	2.5	2.48	215
D	8	92	2.4	2.6	2.58	224

Mix design D will require a spread rate greater than the 220 lbm/yd^2 estimated in the plans.

The answer is (D).